MATHEMATICS
IN THE SOCIAL SCIENCES
AND OTHER ESSAYS

MATHEMATICS
IN THE
SOCIAL SCIENCES
AND OTHER ESSAYS

by

RICHARD STONE

CHAPMAN AND HALL LTD
11 NEW FETTER LANE LONDON EC4

First published 1966
© Richard Stone 1966
Printed in Great Britain by
Spottiswoode, Ballantyne & Co. Ltd.
London and Colchester

FOREWORD

That mathematics is an indispensable tool in the study of the social sciences is no longer a very controversial statement. It is now generally agreed that mathematical methods are necessary both at the theoretical level, to formulate problems precisely, to draw conclusions from postulates and to gain insight into the workings of complicated processes, and at the applied level, to measure variables, to estimate parameters and to organize the elaborate calculations involved in reaching empirical results. The seventeen essays collected in this book are all illustrations of my belief in this principle.

But there is another, perhaps less widely recognized side to the service-ableness of mathematics in the social sciences, which I hope will also be apparent to the reader of these essays: namely, that techniques generated for some specific purpose in one science can quite often be fruitfully applied in another. Thus the mathematical formulation of the birth and death process, originally devised to investigate change in living popu-lations, can be applied to populations of inanimate objects such as telegraph poles or vehicles; the epidemic model, developed primarily to describe the spread of infectious diseases, can be applied to other infectious phe-nomena such as the wish for higher education or the demand for a new product; input–output analysis, evolved in the first place to study industrial interdependence, can be adapted to the study of demographic flows; and so on. All of which suggests that there is room in the curricula of the social sciences for courses designed to bring together a variety of techniques at present usually regarded as belonging to separate disciplines.

And this leads on to a third and equally important point. Since I am an economist, most of my examples relate to economics, and more particularly to the computable model of the British economy which has been my main concern in recent years. My work on this model has brought home to me, even more forcibly than my earlier work had, the difficulty of disengaging the economic aspects of life from their demographic, social and psycho-logical setting. It is not chance or whim that often prompts the builder of large economic models to study population movements, educational arrangements, attitudes to innovation; it is the realization that what he is really modelling is the whole socio-economic process. At the theoretical level the different facets of this process can and should be studied in iso-lation; but at the applied level such specialized studies will not yield their full benefit to society until we have learnt to connect them.

In concluding this foreword I should like to thank the editors of the various journals, conference volumes etc., most of them foreign, in which these essays were first published. The essays are reprinted substantially as they appeared originally except that the first has been expanded and in a few others some sections have been rewritten to avoid unnecessary overlap.

Cambridge   RICHARD STONE
October 1965

PLACE AND DATE OF ORIGINAL PUBLICATION

I   *The Scientific American*, vol. 211, no. 3, 1964.

II   *L'Industria*, vol. 4, no. 4, 1963.

III   *L'Industria*, vol. 3, no. 4, 1962.

IV   *Economia Internazionale*, vol. VIII, no. 1, 1955.

V   *Logic, Methodology and Philosophy of Science: Proceedings of the 1960 International Congress*. Stanford University Press, 1962.

VI   *Operational Research Quarterly*, vol. 14, no. 1, 1963.

VII   *The Manchester School of Economic and Social Studies*, vol. XXX, no. 1, 1962.

VIII   *On Political Economy and Econometrics* (in honour of Oskar Lange). Polish Scientific Publishers, Warsaw, 1964.

IX   *Minerva*, vol. III, no. 2, 1965.

X   *Regional Economic Planning: Techniques of Analysis*. European Productivity Agency of the O.E.E.C., Paris, 1961.

XI   *Journal of Regional Science*, vol. 2, no. 2, 1960.

XII   *Przeglad Statystyczny*, vol. 8, no. 2, 1961.

XIII   *Przeglad Statystyczny*, vol. 7, no. 3, 1960.

XIV   *Problems of Economic Dynamics and Planning* (in honour of Michal Kalecki). Polish Scientific Publishers, Warsaw, 1964.

XV   *The Manchester School of Economic and Social Studies*, vol. XXXII, no. 2, 1964.

XVI   *Bulletin of the International Statistical Institute*, vol. XXXIX, pt. 3, 1962.

XVII   *Econometric Analysis for National Economic Planning*. Butterworths, London, 1964.

CONTENTS

I

MATHEMATICS IN THE SOCIAL SCIENCES

1. INTRODUCTION

Except in a few obstinate pockets of resistance, the use of mathematics in the social sciences is now generally accepted. The reason is not to be found in the outcome of any high-flown philosophical battle but in a number of simple facts. In the first place, many branches of the social sciences are obviously, one might almost say aggressively, quantitative; demography and economics are clear examples of this. In the second place, while theories about the complex systems which are the subject matter of the social sciences can be expressed verbally, their analysis and comparison are greatly helped by formulating them mathematically. In the third place, the application of such theories must remain very general unless the terms in their relationships can be quantified. In the fourth place, mathematics provides a means of obtaining insight even into subjects whose concepts are rather vague and where precise information is hard to come by. Finally, in the social sciences we are interested not only in a description of what happens and of how the different parts of the social system are related, but also in the rational processes that lie behind effective as opposed to ineffective decisions; to a large extent these processes too can be formulated and analysed mathematically, so that our decisions may eventually come to rest a little more on knowledge and a little less on guesswork than they do at present.

Now that the possibility of expressing mathematically concepts which had hitherto been thought of only verbally has been demonstrated, and the emotional resistance to the use of mathematics has been largely overcome, more and more scholars are realizing the value of applying mathematical methods to their branch of the social sciences. As things stand at the moment, the advances achieved are very uneven and the unity of the subject falls short of what one would like and expect. But the mathematical movement is now fairly launched and in the next generation we shall probably see an integration of studies in many different fields which are at present the preserves of distinct and often warring specialisms.

Some seventy-five years ago, when the great American economist Irving Fisher wrote his doctoral dissertation he considered that there were in the whole world only about fifty books and articles on mathematical economics which were worthy of the name. Now the position is very different: not only in economics but in all the social sciences mathematical books and articles appear by the thousand every year. So in a short paper like this one a complete survey of the subject is out of question; the best I feel I can do is to give a number of examples set out under broad headings. There is nothing final or necessary about these headings: they simply provide one way of organizing a very large subject.

2. STRUCTURE AND CHANGE

It is obviously desirable to be clear about the structure of anything we are studying, about the relationships which connect its various parts and about the factors which make each part, and hence the whole structure, change. In the social sciences, especially where empirical analysis is concerned, much use is made of finite mathematics, and in particular of matrices, matrix algebra and difference equations, since they correspond in an obvious way to the discrete observations that form the basis of most empirical work. This does not mean that traditional techniques such as the infinitesimal calculus, and in particular differential equations, do not also have many advantages, especially where purely theoretical analysis is concerned. In this section I shall give some examples of the application of matrix algebra and of differential equations to the description and analysis of systems, whether these systems belong to the field of demography, anthropology, sociology or economics. Thus I hope to bring out a certain unity in the structure of many apparently very different problems, a certain sense of having been there before as we pursue the various techniques from one science to another.

(a) *Births and survivals.* Suppose we want to study the structure and probable development of some population. For this we need three sets of data: (i) the numbers of people of different ages alive at a certain date; (ii) the numbers surviving over a certain time-interval, say a year, immediately following that date; and (iii) the numbers born during that same time-interval to people of different ages. The information about the age composition of the population can be set out in a vector, or single column of figures: if we take the year as our time unit, each element in this vector will record the number of people who at the date in question had reached a particular year of age, 0, 1, 2, and so on. The information about births and survivals can be arranged in a square matrix in which each row and each column will relate, again, to a particular age-group: the number of children born to people in each age-group will appear in the first row of the matrix (I shall not go here into the complications arising from the fact that husbands and wives often belong to different age-groups); and the numbers of people surviving from one age to the next will appear in the diagonal immediately below the leading diagonal which runs from the top left-hand to the bottom right-hand corner of the matrix. All other elements in the matrix will be zero.

Thus we have a vector which shows us at a glance the age composition of the population at the beginning of the period and a matrix which shows us the pattern of births and survivals in this population during the period. Now, by dividing the births and survivals in each column of the matrix by the number of people in the corresponding age-group in the vector, we can form a coefficient matrix of birth rates and survival rates. If we then pre-multiply the vector by the coefficient matrix we shall obtain another vector, which will be an estimate of the population structure at the end of the period. Similarly, we can obtain an estimate of the population structure θ years from now by premultiplying the original vector by the coefficient matrix raised to the power of θ. This estimate, of course, will be correct only if the coefficients are fixed, that is if the pattern of births and survivals stays constant over the θ years, so that the coefficient matrix retains its validity in the future. A population whose structure and growth are characterized by a fixed coefficient matrix will eventually reach a stable age composition and a constant rate of growth.

In practice, however, the coefficients may not be fixed: they may be subject either to chance influences or to systematic influences or to both. It is not difficult to work out the conditions under which a population would tend to an upper (or lower) limit [97, 98]; and, by complicating the model, many problems of ecology and epidemiology can be studied [9, 13].

So far I have identified populations with human populations, but a precisely similar set-up can be used in the study of populations of industrial equipment, such as telegraph poles or railroad cars, and thus help to solve some of the problems of industrial renewal [99, 144]. In this case, specific birth rates are replaced by investment rates and, whatever the age composition of the initial stock, it is possible to work out the replacements and extensions needed to keep the stock on any given time-path.

(b) *Marriage and descent.* The study of certain types of tribal kinship can be helped by the use of a pair of permutation matrices, that is square matrices which have a single 1 in each row and each column and 0's everywhere else. In these studies, the society in question is divided into clans and each clan is associated with a row-and-column pair in each of the matrices. In the marriage matrix, a 1 at the intersection of row j and column k, say, signifies that a man of clan j may marry a woman of clan k. In the descent matrix, a 1 at the intersection of row j and column k signifies that the children of a man of clan j are members of clan k. Postmultiplication of the marriage by the descent matrix gives rise to another permutation matrix, which shows the clan to which the children of a man's wife's brother belong. More generally, the product of any ordered sequence of the two matrices will show the clan of some relative of a man in each clan.

In theory the number of possibilities is very large, but if we use our anthropological knowledge to specify the characteristics of the two original matrices for a given restriction on their size, we find that we can classify societies according to their kinship structure and, further, that actual societies exist which conform in greater or less degree to these ideal types [188].

Incidentally, the set of kinship matrices generated by the marriage and descent matrices constitute an abstract group, and their connection with one another can be set out in the form of a multiplication table.

(*c*) *Class structure and social mobility*. There are many other uses to which marriage and descent matrices can be put. For example, we may be interested in the stability of a society's class structure. Thus let us consider a society of four classes in which: (i) at least one member of any marriage must belong to class 4; (ii) if the father is of class 4, then the child will be of the class of its mother; and (iii) if the mother is of class 4, then the child will be of the class immediately below that of its father, except that if the father is also of class 4, then the child will be of class 4. If in such a society each individual marries once and each marriage produces one child of each sex, we find that a stable class structure is possible only if the first two classes are initially absent from the population [64]. This means that a stable class structure is in fact impossible in that society, since otherwise we should have spoken of two, and not of four, classes in the first place.

In modern societies, kinship patterns may be of relatively little significance but we may nevertheless be interested in social class patterns defined by more numerous criteria [63]. For this purpose sociologists provide us with square matrices in which the row-and-column pairs are associated, again, with a given class, but in which the entry at the intersection of row j and column k measures the probability that the son of a father in class k will himself move into class j. If we abstract from questions of differential fertility between classes, we can find out from such matrices the class structure to which a society will tend if it continues to recruit new members to its different classes in the same way as it does at present [125, 126]. From the same basic data it is possible to calculate the number of generations spent on average in any particular class and to compare this with what would happen on some definition of a perfectly mobile society.

(*d*) *Group interaction*. In sociology we frequently have to deal with concepts which may seem particularly elusive and intractable, such as the behaviour of individuals when brought together in a group. In such cases mathematics often helps us at least to state the problem clearly [142]. Thus we might characterize the behaviour of a social group by four variables: the intensity of interaction among members; the level of friendliness among members; the amount of activity among members; and the amount of activity imposed on members by their environment. These four variables can be related as follows: interaction is a linear compound of friendliness and internal activity; the rate of change of friendliness through time is proportional to the difference between interaction and some multiple of friendliness; and the rate of change of internal activity through time is a linear compound of (i) the excess of friendliness over some multiple of internal activity, and (ii) the excess of externally imposed activity over internal activity. These relationships constitute a set of linear differential equations with constant coefficients, which can be solved completely and explicitly. This model can be used to study the conditions of equilibrium and stability in the group and to see how the group would behave if it were disturbed from a position of stable equilibrium.

(*e*) *Social accounting*. Economists have the problem of reducing innumerable transactions concerned with production, consumption, accumulation and foreign trade to some kind of order, and of providing a framework to contain the estimates for any particular economy [154, 162]. The answer to this problem is to view the world economy as a vast system of interlocking accounts. This accounting system can then be set out in the form of a square matrix in which the row-and-column pairs each represent an account, incomings (or revenues) being shown in the rows and outgoings (or costs) being shown in the columns. Since each account must balance, the sum of the entries in any row is equal to the sum of the entries in the corresponding column. Since the accounting system is closed, its entries are connected by a number of identities equal to one less than the total number of accounts.

If we tried to record in detail in a single matrix all the flows in an economic system, the resulting matrix would be impossibly large and unmanageable. We must therefore consolidate the accounts into convenient classes, choosing our classification according to which aspect of the economy we want to study. Consolidation is achieved by: (i) partitioning the matrix conformably into a number of submatrices; (ii) adding up the entries within each submatrix; and (iii) discarding the resulting diagonal elements, which represent intra-class transactions. Two forms of partitioning are especially useful: that which leads to the national accounts and that which leads to inter-industry, or input–output, accounts. Both are described below.

(*f*) *The national accounts and a model of production growth*. In order to form the national accounts we must partition the social accounting matrix into four classes of accounts, one for production, one for consumption, one for accumulation and one for the rest of the world. In this way all the transactions in the economy are reduced to a very small number of large aggregates.

For some purposes it is useful to simplify matters still further and to consider a closed economy, where the four accounts are reduced to three by the disappearance of the foreign account. In such an economy, the production account receives money from the consumption account in exchange for consumption goods and services, and from the accumulation account in exchange for investment goods, and pays out the value of its sales in the form of income. This income is the only receipt into the consumption account, where it is divided between expenditure on consumption goods, which flows back into the production account, and saving, which flows into the accumulation account. Finally the accumulation account, having received this saving, pays it into the production account in return for investment goods, thus closing the circuit.

This model has four relationships. The first is an accounting identity which states that total income (or product) is equal to the sum of expenditure on consumption goods and saving. The second is also an accounting identity, which states that saving is equal to expenditure on investment goods. The third relates to the behaviour of savers and states that people save a fixed proportion, α say, of their income. The fourth relates to the technical conditions of production and states that there is a fixed coefficient of proportionality, β say, between additional capital, or expenditure on investment goods, and additional production.

An economy governed by these relationships will grow at a constant rate of α/β, which means that in order to grow faster it must either save a larger proportion of its income, that is increase α, or learn to use less capital per unit of output, that is reduce β. For example, if it saves 10 per cent of its income and has a capital–output ratio of 2·5, it will grow at a rate of 4 per cent; if it could either increase its saving to 12·5 per cent or lower its capital–output ratio to 2, it would grow at a rate of 5 per cent.

(*g*) *Industrial inputs and outputs.* The national accounts reduce the economic transactions of a country to their simplest terms. If we want to analyse the productive system in detail, we must deconsolidate the national account for production into a number of industries [151]. At the same time, for simplicity, we can consolidate all the non-production accounts into one. The resulting set of accounts can be presented as a partitioned matrix consisting of: (i) a square submatrix in which each row-and-column pair corresponds to an industry and whose entries show the raw materials, fuels, etc., that is the intermediate inputs and outputs, flowing from one industry to another within the productive system; (ii) a column vector whose entries show the amounts of product which the various industries deliver outside the productive system, that is the industries' final outputs; (iii) a row vector, whose entries show the inputs into the productive system which are not produced by the system itself, that is the inputs of labour and capital, or primary inputs, into the industries; and (iv) a single zero, which appears at the intersection of these two vectors in the bottom right-hand corner of the matrix and indicates that the non-production accounts are consolidated.

If we divide the entries in each column of the input–output submatrix by the total output of the corresponding industry, we obtain a matrix of coefficients in which the entry at the intersection of row j and column k measures the amount of product j needed directly to produce one unit of product k. If we subtract this coefficient matrix from the unit matrix, that is from a matrix which has 1's in the leading diagonal and 0's in all its other cells, and take the inverse, we obtain a new matrix, whose elements measure the amount of product j needed directly and indirectly to produce one unit of output k. Thus if we can foresee what the vector of final outputs, or final demands, will be at some future date, we can estimate the total output levels to which these demands will give rise in the various industries.

As in the case of the birth and survival model, this estimate will be correct only if the coefficients are fixed, that is if the techniques of production do not change. In practice the technology of production does change, and such changes can be introduced into the input–output model, although I shall not go here into the methods used to do so [32, 35].

(*h*) *Education and training*. The educational system, by which I mean schools, universities and all forms of professional and industrial training, can be studied with the aid of an educational matrix not dissimilar from the industrial input–output matrix, as described in [IX] below. We can regard the different stages of the educational system as so many industries, or processes, through which the students pass, first as raw materials, then as semi-finished products, and eventually as final products, or graduates. For any individual, graduation takes place when he passes out of the system, whatever stage he may have reached at the time.

If we can form an opinion on the composition of graduates that we shall need at some future date, we can work out what this means in terms of the activity levels of the different educational processes, essentially by combining input–output analysis with an analysis of the future population structure. Unlike the static input–output model described above, however, the educational model is formulated in dynamic terms, so that the activity levels of the various processes in each year must be such as to provide not only the graduates required at the end of that year but also the student inputs into the various processes for the following year.

(*i*) *Teachers and graduates*. If we are concerned with the future supply of teachers required to extend an educational programme, we need to know how teachers and graduates interact. Let us consider the simplest possible case: a single educational process in which the number of graduates is proportional to the number of teachers. Let us denote the graduate–teacher ratio by α. The net rate of inflow of teachers is made up of two terms: a supposedly constant proportion of graduates, β say, who become teachers; and a supposedly constant proportion of the existing stock of teachers, γ say, who are lost to teaching through change of job, retirement or death. From these two relationships it is not very difficult to see that the net rate of growth of teachers and of graduates, δ say, is $\alpha\beta - \gamma$. Thus if we want an increase in the rate of growth of graduates, we must either: (i) get the existing number of teachers to turn out more graduates, that is increase α; or (ii) persuade more graduates to become teachers, that is increase β; or (iii) diminish the wastage rate in the stock of teachers, that is reduce γ.

These relationships can be illustrated geometrically. If we measure α and β on two rectangular axes and draw in a family of rectangular hyperbolae, each member of this family will correspond to a different gross rate of growth, $\gamma + \delta$, in the stock of teachers; that is, it will show all possible combinations of α and β that will yield this rate.

Even supposing that there is nothing much we can do about γ, there are doubtless many things we could do to influence α and β. One of these might be to pay teachers better. We could easily imagine that both α and β were related hyperbolically to pay, in such a way that at very low rates of pay there would be no work and no inflow of graduates into teaching, while at very high rates of pay both the work done and the inflow of graduates would approach upper limits. In this case it is easy to show, by eliminating the rate of pay from the two equations for α and for β, that the relationship between α and β is linear. This linear relationship can be inserted in the geometrical construction described above. By its intersection with the various members of the family of rectangular hyperbolae, it will determine in each case values of α and β and enable the corresponding rates of pay to be calculated.

3. DECISIONS

So far I have talked about describing the world as it is, without considering the mechanism which determines how it comes to be that way. To some extent this mechanism may lie outside the social sciences, in biology, in individual psychology, in technology: all of which the social sciences may reasonably take as data. But to a large extent the maintenance and modification of social patterns depend on innumerable private and public decisions, consciously undertaken to achieve certain ends. In the social sciences, therefore, the study of the decision process is an essential element.

This large and complex topic can best be approached by distinguishing first between decisions taken under conditions of certainty and decisions taken under conditions of uncertainty, and then, within these two main categories, between decisions that relate to a single period and decisions that relate to a course of action over several periods. In this section I shall describe some of the techniques available to help the decision-maker in these different situations: programming, games theory, statistical decision theory and so on. Here again we find a number of new techniques which lean heavily on the methods of finite mathematics, although, as we shall see, many decision problems are of a kind that can be solved, theoretically at least, by the older methods of undetermined multipliers and of the calculus of variations.

(*a*) *Single-stage decisions under certainty.* A good example to begin with is the theory of consumers' behaviour, developed in the late nineteenth century. According to this theory [70, 168, 190], a consumer has a well-defined system of preferences, and the utility, or satisfaction, which he derives from what he buys depends on the quantities of the various goods and services he buys. He has a fixed amount of money to spend, his income, and is faced with a fixed set of prices. His aim is to maximize his utility subject to the constraint set by his income and by the price system.

This is a constrained-maximum problem which can be solved by the method of undetermined multipliers and admits of a simple geometrical interpretation. Consider an Euclidean space of *n* dimensions, along whose axes the quantities of the different goods and services are measured. In the wholly positive part of this space, consider an infinite set of non-intersecting hyper-surfaces which have continuously turning tangents and are strictly convex to the origin. The consumer is indifferent between the various collections of goods represented by any point on a given hyper-surface, but prefers to be on a hyper-surface as remote from the origin as possible. His income and the prices determine a plane in the space, which divides it into two regions: an attainable one, towards the origin, and an unattainable one, outwards from the origin. The hyper-surface which is the highest he can attain will have a point in common with the constraining plane and this common point defines the most desirable set of goods that he can afford.

The outcome of this theory is that, for given preferences, the demand for each good is a function of income and of the price system. If the form of the preference system, or utility function, is assumed, it may be possible to infer the form of the demand functions connecting the amounts of the different goods demanded with income and prices. For example, a hyperbolic utility function corresponds to demand functions in which expenditure on each commodity is a linear homogeneous function of income and of all the prices [60], or, put in another way, the consumer buys certain quantities of each commodity, corresponding to his concept of his standard of living, and then spreads any money that remains over the various goods in fixed proportions.

The principal use of this theory is to make statements about aggregate or market demands. This can be done by summing individual demand functions over a community of consumers and replacing the innumerable individual incomes that appear in the aggregate function by one or more moments of the income distribution. In this aggregated form the theory can be tested against observations, and experience shows that it provides a good first approximation to consumers' behaviour in the real world [168, XIV].

But it needs elaboration: preferences change systematically; people take time to adapt themselves to changing circumstances; people are influenced by what other people do. A particularly interesting type of adaptation is connected with the introduction of new goods. The response to a new good involves a learning process and in many respects resembles an epidemic. The rate of acceptance depends in part on the number of people who already possess the new good and in part on the number of people who do not possess it. This learning process can be well approximated by a positively skewed curve, say the log-normal integral [10, 11].

It commonly happens that a decision problem, though somewhat similar to that just described, does not admit of a solution by the classical method, because the function to be maximized (or minimized) is linear. An example of this is the minimum-cost-diet problem, for the solution of which we must turn to the modern technique of linear programming [50]. This problem can be stated as follows. An adequate diet is defined in terms of certain minimum quantities of nutritional elements: calories, proteins, vitamins, etc. A variety of foods are available which contain these nutritional elements in known quantities and whose prices are fixed. The question is: how much of each food should one buy in order to meet the nutritional requirements at minimum cost. This is equivalent to saying: minimize the sum of the unknown quantities of the various foods each multiplied by its known price, subject to a number of inequalities. These inequalities state, first, that the collection of quantities bought must provide enough of each nutritional element and, second, that the amount bought of each food must be non-negative.

Again, this problem and its solution can be interpreted geometrically. I shall take the simple case of two foods and three nutritional elements. Let the quantities of the two foods be measured along two rectangular axes. On each axis mark off the amount of one of the foods which by itself would provide the minimum quantity needed of one of the elements, and then join the points so determined to form a straight line in the positive quadrant. Do the same for each of the other nutritional elements. If all the constraints are somewhere effective, these lines will cross, and we can join segments of the straight lines to show the boundary along which all the constraints are satisfied. This boundary will be convex to the origin. Now let us introduce the ratio of the prices of the two goods. We can do this by marking off on the axes the quantities of each food that can be bought for, say, £1 and by joining these two points by a straight line. If we slide this line in and out, always parallel to itself, we shall discover the position nearest to the origin at which it has one or more points in common with the convex boundary. Any of these points defines a combination of the two foods that provides a solution to the problem With *n* foods the same construction is repeated in *n* dimensions, and as *n* increases, the number of calculations needed to find a solution increases too.

In practice, the solution of this particular problem usually turns out to involve very few foods and to offer a pretty unacceptable diet. The reason is to be found in the very narrow formulation of the problem. For example, no account is taken of the bulk of the food that constitutes the minimum-cost diet. This always turns out to be very large. As has recently been found, a much more varied diet emerges, with no appreciable increase in cost, if a limit is put on the weight of the food to be consumed each day. This illustrates a characteristic feature of all complex calculations: mathematical methods are literal methods, they solve problems as they are formulated. It is up to the investigator to see that they are formulated sensibly, although it must be admitted that when it comes to the point it is often very difficult to do this.

(b) *Multi-stage decisions under certainty*. Often when taking a decision we have to consider what to do so that some variable will have a desirable time-path. In such cases what we have to maximize (or minimize) is a functional rather than a function. The classical technique for this purpose is the calculus of variations. Let us, for instance, ask the question: what is the optimum rate of saving for a society? If a society saves very little it can spend correspondingly more on consumption. In the short run, therefore, it can enjoy a relatively high standard of living; but in the long run, since it is adding very little to its capital equipment, it cannot expect this standard to rise very much. So we must ask another question: what would this society be willing to give up now in order to obtain so much more in the future? And we must imagine this question being asked continuously into a future indefinitely prolonged.

If we take the case of a simple economy with only one product that can either be consumed or added to the capital stock, we can formulate this problem as follows. The output of the good is a function of labour and capital inputs; to make matters simpler still, let us assume that the population and the labour force are both constant over time. The amount of the good that we add to the capital stock at any period is equal to the amount produced less the amount consumed. The satisfaction that society derives from its economic activity is equal to the utility of consumption (which is assumed to be a function of the amount of the good consumed) less the disutility, or dislike, of work (which is a function of the amount of work performed). We should like to maximize this satisfaction over an indefinitely prolonged future.

This problem can be solved by means of the calculus of variations provided we assume that there is an upper limit to the satisfaction which society can derive from its economic activity, however much it consumes and however little work it does [131, V]. On this assumption we can calculate the optimum rate of saving at any time by: (i) subtracting the satisfaction currently enjoyed from the maximum satisfaction; and (ii) dividing this difference by the marginal utility of consumption, that is by the rate at which the utility derived from consumption changes as consumption itself increases.

This may appear to be an unexpectedly strong conclusion: we seem to have got a lot out without putting very much in. This is a good illustration of the insight that can be gained into a very complex situation by the application of mathematics to the central features of the problem.

But for practical purposes we need rather more realism, and as we elaborate the formulation of such problems in the interests of realism we usually find either that we exceed the scope of the classical method or that we are forced into obvious artificialities. Fortunately in recent years methods such as dynamic programming [16, 17] and the maximum principle [124] have been developed, which provide a computational approach to problems outside the range of classical methods.

(*c*) *Decisions under uncertainty.* At this point so many techniques crowd together that I shall have to adopt a less extensive treatment. Let us consider: first, uncertainty due to uncontrolled events; second, uncertainty due to the behaviour of others in a conflict situation; and third, the techniques already discussed when certainty is replaced by uncertainty.

Compared with my earlier examples, the effect of uncertainty is to replace a supposedly known magnitude by a distribution of magnitudes, and our problem now is to find out the nature of this distribution and to decide how it should influence our decision. Such questions take us into the field of probability and statistics.

A simple example can be taken from gambling. If we are dealing with ordinary coins, we may reasonably assume, from generations of experience, that the probability of obtaining a head at a single throw is for all practical purposes one-half. If we are asked what the chance is of obtaining a head at each of two throws, we shall answer without much hesitation that it is one-quarter; and more generally, that the chance of obtaining a head at each of n throws is equal to one-half raised to the power of n.

This apparently obvious result is due to our knowledge of the property of coins. But we might run into a situation in which the coins were biased, that is in which the probability of obtaining a head at a single throw was different from one-half. In such a case we should lose money until we found out what was going on. In fact, we might run into an even more complicated situation in which the coins were not all alike but exhibited different degrees of bias. If, for example, the probability of obtaining a head at a single throw ranged evenly between the limits of 0 and 1, it is not difficult to show (II) that the chance of obtaining a head at each of two throws is not one-quarter but one-third. This, in fact, is a celebrated solution of the problem which did nothing to enhance its author's reputation among probability theorists; yet from a practical point of view it is in a way more interesting than the orthodox solution. Although it lacks generality in that it assumes a particular distribution for the probability of obtaining a head at a single throw, it at least draws our attention to the fact that our solution depends on this distribution. Modern statistical decision theory [130, 139] shows how dangerous it may be to make simple assumptions about such distributions in the absence of knowledge, and emphasizes the importance of spending money to obtain such knowledge if the amount at stake warrants it.

There is another aspect of decisions that needs to be taken into account: an individual may be more concerned to avoid a large loss than to achieve a large gain. Thus, in a business context, he may be willing to incur extra costs in holding stocks of some good in order not to keep customers waiting, not because the immediate balance of advantage is equal but because he does not want to get a reputation for long delivery periods which in the end may take his customers away from him. In other words, just as the decision-maker must try to form an accurate estimate of objective probabilities, so must he also try to form an accurate estimate of subjective valuations which are relevant to him. This brings me to my second topic in this section: uncertainties arising from conflict and coalitions. This topic is the subject matter of another new technique, the theory of games [117, 140, 141].

The players in a game of strategy do not know exactly what moves the other players will make, yet their own moves depend on the responses they expect from the others. At least in simple cases, we can work out the pure strategies available to the players and the results which flow from any particular combination of strategies.

In the case of a two-person game in which one player loses what the other gains, these results can be presented in the form of a matrix in which the entry at the intersection of row j and column k indicates the gain to the first player (and therefore the loss to the second player) if the first player adopts strategy j and his opponent adopts strategy k. If we form a column vector of row minima from this matrix, we obtain the minimum amounts that the first player can be sure of gaining for each of his own strategies. If we form a row vector of column maxima, we obtain the maximum amounts that the first player can be sure of gaining for each of his opponent's strategies. A sensible course of action for the first player would be to choose a strategy such that his minimum gain was as large as possible whatever his opponent did; and a sensible course of action for the second player would be to choose a strategy which would minimize the maximum amount that his opponent could extract from him. If the highest minimum value for the first player is equal to the lowest maximum value for his opponent, the strategies which it will pay the players to adopt are determinate, and neither player would have any incentive to change his strategy even if he knew what his opponent's strategy was.

If we consider only pure strategies, games are not necessarily determinate; but if we are willing to consider mixed strategies, that is to say the use of different pure strategies with preassigned probabilities, then it can be shown that any zero-sum two-person game with a finite number of pure strategies can be rendered determinate.

When more than two players are involved, an additional complication arises: the players can form coalitions. One of the objects of the theory of games is to show the circumstances in which coalitions are likely to be formed and prove stable. The number of possibilities soon becomes formidable [147], and this fact goes a long way to account for the relative simplicity of economic theories based on the assumption of perfect competition or perfect monopoly compared with those based on the intermediate assumptions of oligopoly or imperfect competition.

It is an interesting fact that many of the theories and techniques I have been discussing, though superficially different, turn out on closer inspection to have many points in common [100]. Thus there is a mathematical equivalence between linear programming and finding a solution to a game; and there is an intimate connection between the theory of games and the methods used in statistical decision theory.

The final point I want to make here is that the programming methods which I described when discussing decisions under certainty can in some cases be extended to study situations where there is uncertainty either in the constraints or in the preference function [169]. The uncertainty is assumed to be of a probabilistic kind, and it can be shown that in a wide class of cases the decision which maximizes expected utility is the same as that which maximizes the preference function subject to the condition that the uncertain coefficients assume their expected values. Even where a complete solution cannot be reached, it is sometimes possible to gain an insight into the effects of uncertainty on the outcome of decisions.

4. SYSTEMS ANALYSIS

In this section I shall discuss briefly some of the problems of modelling the whole socio-economic system, an undertaking that brings into play all the preceding techniques, and shall try to show the role of models in real life, where we are interested both in understanding how the system works and in ensuring that it works as well as possible.

In order to analyse a system we need a model which sets out the variables in the system and shows how they are related. To establish the variables, we need not only observation and measurement, but also taxonomic work of the kind I indicated in my discussion of social accounting. To establish the relationships, we need not only theories formulated mathematically but also statistical techniques which will enable us to estimate the numerous parameters contained in the relationships, partly from observations of the past and partly from expectations about changes in the future. The theories we use may vary in depth. Some may represent nothing more than a formalization of observed regularities without any obvious rationale.

Others, such as the theory of consumers' behaviour described in section 3 (*a*), may be a rational explanation of some accepted mechanism of decision, in which case the mechanism of decision is automatically built into the model and the results to which it leads are treated as part of the objective situation.

The main purpose of such a model is to work out the implications of any assumptions we choose to make, and to do this in a way that ensures consistency and, as far as possible, realism. Let us suppose that we want merely to find out how the system is likely to develop 'naturally' in the future. In this case, by matching the number of variables and relationships, the model can be made closed and determinate; and by formulating the relationships dynamically, it can be used to trace out the future paths of the variables from a given set of initial values, just as the little model of saving and investment in section 2 (*f*) did. But, as that example also illustrated, we may be interested in trying to discover how to bring about a state of affairs which does not seem to come about naturally: how to reduce unemployment, say, or increase the proportion of higher graduates, or avoid recurrent balance of payments crises. In this case we must open the model at some point and introduce into it a new feature: a precise statement of aims. The model can now be used to see how the system could be balanced out on the assumption that these aims were met.

Thus, just as we obtained a model by combining theories and facts, so we can obtain a policy by combining a model with a set of aims. Attempts to state our aims are likely to lead us back to utility theory and to the comparison of social costs and benefits, a comparison which requires not only valuations but also the analysis of the interaction of the variables to be valued, as indicated by the model.

But this is not the end of the story. Whether decision processes are considered simply as part of the objective situation (as in the extreme theory of *laissez faire*) or as modifiable to some extent by public action (as in all communities of the real world), the system may not keep on an even course, either through an inherent tendency to oscillate or through a poor capacity to recover from the succession of shocks to which it is inevitably subject. Socio-economic systems, like their biological and engineering counterparts, possess automatic control mechanisms; an example of this in economics is the price mechanism. Often, however, these mechanisms do not function very well, partly because they are implicitly based on limited aims and partly because they work with limited information. This is why in every country efforts are made, in greater or less degree, to design devices which will improve them. But whether we are content with purely automatic controls or try to supplement these by conscious intervention, the fact is that controls exist and, when combined with a policy, give rise to a plan of action.

This plan in turn combines with events to give us our experience of socio-economic life. This experience feeds back to modify the theories we accept, the facts we consider relevant, the aims that appeal to us and the controls we regard as efficient. By modifying these factors, experience modifies our models, policies and plans. And so on, until we may hope some day to live in a world which functions a little better and is less at the mercy of events.

Of all the social sciences, that in which the technique of macro-models is perhaps most advanced is economics. Indeed, large quantitative models of the economy, intended for practical use, are at present being built in many countries, irrespective of their political complexion [36]. These models are made possible by the recent development of the electronic computer; a generation ago, when econometrics was coming into being with the help of desk calculators, they would have been unthinkable.

At this point I must give up any attempt to illustrate my remarks with several examples. I shall therefore concentrate on a single one: the computable model of the British economy which my colleagues and I are working on at Cambridge. I cannot hope in a few words to explain this model in detail, but I shall try to show how mathematics comes into the picture at every turn and how any such model must inevitably develop into something more than a model of the economic system pure and simple. Many aspects of this model are described more fully in later essays in this volume and the model as a whole is outlined in [XVII]. A complete account of our work is gradually appearing in the series *A Programme for Growth* [30 to 36].

Our model is designed to help in analysing the conditions and consequences of realizing in the future certain economic aims, such as faster economic growth. Our first task has been to construct a model of a steady state of growth starting from a future year, which for the moment we have set at 1970. Our second task, on which we have begun work only recently, is to study the transient problems of realizing that state given the present state of the economy. Eventually these two parts of the model will be made to interact.

The variables in the model are the entries in a social accounting matrix and the prices and quantities associated with these entries. At present this matrix contains 253 row-and-column pairs, each of which represents a balancing account. By working within this framework we can at least be sure that our results are consistent in an arithmetical and accounting sense. This matrix is an extension of the one described below in [XVI].

In order to set the model to work, certain basic assumptions are made at the outset and their implications are then spelt out by means of economic relationships. For example, total private consumption is assumed to have reached a certain level by 1970 and to grow thenceforth at a certain pre-assigned rate; demand relationships are then used to allocate the total among its constituent goods and services. A number of other components of final demand, such as government consumption and exports, are treated in a similar way. Production and foreign trade relationships are used to work out requirements for industrial equipment and stocks and for intermediate product, and to divide the total demand for each product group between domestic and foreign supplies.

The numerical inputs (parameters and conditions) needed for a computer-run to obtain these supply-and-demand balances number between five and six thousand. A run involves about thirty million multiplications and takes twenty-two seconds on the Atlas computer.

The model is in a continuous state of improvement and extension, and this means that the computing programme needs to be flexible. To this end we have written the programme in stages, so that one stage can be changed without affecting the others; we have made extensive use of matrix methods, so that as the size of the model increases, all we need do is change some of the input parameters; and we have taken full advantage of the computer's ability to handle iterative and relaxation processes. By such means we try to overcome the endless problems set by revisions of data, alterations in the form of the relationships and changes in the methods used to estimate the parameters.

The fact that preferences and technology are not fixed faces the economic model-builder with one of his most difficult problems. He is bound to start with a statistical examination of past observations. At the same time he must be ready to recognize change and to bring past relationships up to date and project them into the future. As he does this, he will want to check his results as far as possible against those of people engaged in the various practical activities he is trying to model. Finally, he must be able to absorb into his model the fragmentary but often important information that these people can give him.

Our ideas are moving more and more in the direction of a division of tasks in building economic models. As we increase the size of our original model, we have to decide whether just to add categories to it or to set up a number of sub-models to handle them. On the whole, the second solution seems the more practical. Thus, in addition to the central model, we might have a sub-model for the fuel and power industries, say, another for petro-chemical production, another for financial activities, and so on. The in-dividual models in this model-system would work by means of the exchange of information, leading to convergence in the system as a whole. Such an organization of the work would have several advantages: the task of building and running each sub-model would be decentralized and put in the hands of people with the necessary specific knowledge; the demands of government and business secrecy would be respected; and the modelling would be more realistic than would be possible in a single central model. For example, by means of programming methods we could represent situations involving choice in the individual branches of production, whereas with a single model these situations would be effectively ruled out by the use of input–output relationships fixed in all circumstances.

Although our particular model started with the economic aspects of social life, it has an inevitable tendency to spread. For example, in discussing the parts played by labour, capital and invention in producing goods and services, we need to consider, on the one hand, different skills, and this leads us to study the system of education and training in which these skills are learnt [IX], and, on the other hand, attitudes to research and innovation, and this leads us into the field of social psychology. In the end we shall indeed have to face a complete analysis of the socio-economic system.

II

THE *A PRIORI* AND THE EMPIRICAL IN ECONOMICS

The keynote of my *Plea* is that the work of the economist is 'to disentangle the interwoven effects of complex causes', and that for this general reasoning is essential, but a wide and thorough study of facts is equally essential and that a combination of the two sides of the work is *alone* economics *proper*. Economic theory is in my opinion, as mischievous an imposter when it claims to be economics *proper* as is mere crude unanalysed history.

Six of y^e one, $\frac{1}{2}$ dozen of y^e other!

ALFRED MARSHALL

(in a letter to F. Y. Edgeworth, 28 August 1902)

1. INTRODUCTION

In studying the real world we isolate systems, and apply two methods of investigation. The first consists of making assumptions which can be combined into theories and given a particular form in terms of a theoretical model. The second consists of observing or measuring the components or features of the system so as to obtain a descriptive account of it and can be given the form of a practical specification. The combination of the two methods leads to a model. The purposes of constructing models are: (i) to understand how the system works; (ii) to calculate its outcome in different circumstances; and (iii) to indicate how it should be redesigned so as to function more in accordance with our wishes.

In setting the stage in this way I want to emphasize two features of scientific activity which I think are usually present. First, we are studying systems, that is things composed of interdependent parts. Second, we are interested in the real world and our models must not only satisfy conditions of consistency but also of realism; otherwise we shall only discover how a system of the kind we are studying might operate, not how the actual one we are studying does operate. I should add at this point that we do not begin by building, or even trying to build, perfect models. We begin by modelling the main features of the system we are studying and then try to improve on our prototype. Generally speaking we can never continue this process to its end; the world is so complicated that we must usually content ourselves with stochastic models.

3

2. THE *A PRIORI* AND THE EMPIRICAL

If we examine what we do when we build models, it seems to me that the hard and fast distinction between the *a priori* and the empirical disappears. We build our theoretical models on a slender base of observation and then test them. Similarly our observations and measurements are based to some extent on concepts of what it would be illuminating to observe or measure. This sounds like a hen and egg proposition but that does not matter, because learning is essentially an iterative process. As Pythagoras said, ἀρχὴ δέ τοι ἥμισυ πάντος, and it is irrelevant whether we begin with a theory in search of observations or observations in search of a theory.

It may be worth while to illustrate the proposition that assumptions about the nature of the real world enter into the construction even of very simple models. Consider the question of tossing pennies. Let us assume that we have an unbiased mechanism for tossing pennies and that the *a priori* probability of obtaining a head in a single throw is a number p between 0 and 1. The probability, h say, of obtaining at least one head in n throws, or the probability of not obtaining a tail at each throw, is a function of p and n. According to the generally accepted theory, this function takes the form

$$h = 1 - (1-p)^n \tag{1}$$

With $p = \frac{1}{2}$, that is on the assumption that the penny is unbiased, and $n = 2$, we see that $h = \frac{3}{4}$; with $n = 3$, $h = \frac{7}{8}$, and so on.

This model is based on the implicit assumption that we know enough about pennies to make it worth while to assume a fixed value of p and, in the numerical example, to take a particular value as fixed. But suppose we know nothing about pennies. We might then think that we should allow for an unknown degree of bias, that is we should allow p to range in some way over all values between 0 and 1. If we decided to represent our state of mind by supposing that p might take any value between 0 and 1 with equal probability, then the expression for h would become

$$h = \int_{p=0}^{1} [1 - (1-p)^n] \, dp \tag{2}$$

With $n = 2$, this would yield

$$h = \int_{p=0}^{1} (2p - p^2) \, dp = [p^2 - \tfrac{1}{3}p^3]_0^1 = \tfrac{2}{3} \tag{3}$$

while with $n = 3$ it would yield $\frac{3}{4}$, and so on.

Naturally enough, the two theories yield different results. The difference depends on our knowledge of the properties of pennies. If we assumed very little knowledge as in (2) we should lose a lot of money until experience showed us that a more appropriate model was (1).

It is perhaps interesting to note that the model given by (2) was adopted in preference to (1) by d'Alembert in his article on heads and tails published in 1754 [45] and that he was severely taken to task for this choice by Todhunter [171]. It will also be seen that (2) underlies Carnap's theory of equally likely structures [39]. Indeed, if we are discussing not pennies but mathematical objects, which we might call discs, whose properties are summarized in the distribution of p, then each model is correct for the appropriate distribution. Thus prior empirical knowledge may make a lot of difference to the theory we construct.

In a similar way prior theoretical knowledge may make a lot of difference to what we try to observe and measure. It is not easy to obtain empirical correlates to economic concepts like income and saving. In doing so we must recognize: (i) that such concepts are related by identities of an arithmetical and accounting form; and (ii) that, given this kind of consistency, there are still many variants, some of which are much more useful in analysis than others. If we just go out and drag in observations without any idea of how we shall use them, we are likely to waste our time. This point was made some years ago, in the context of economics, in the debate on measurement without theory [87, 185].

And so, as I said at the beginning, we usually start with theories that have a very small empirical input or with facts that have a very small theoretical input. Once we start, from either end, the *a priori* and the empirical interact. Let us now look at some examples of this interaction in economics.

3. DEMAND THEORY AND CONSUMERS' BEHAVIOUR

The great economist Jevons, writing in the 1870's, laid it down explicitly that the theory of economics must begin with a correct theory of consumption and regarded it as unaccountable and quite paradoxical that English economists should, with a few exceptions, have ignored this doctrine. He rightly considered his theory of utility as a theory of consumption, and his ideas were further developed by his contemporaries and successors until in the works of Pareto we find a fully elaborated theory of consumers' behaviour which makes use of the idea of indifference surfaces. This theory was examined by Wold, some twenty years ago, in a series of papers [189], in which he also demonstrated that other theories that had been propounded were equivalent to it provided that they were rendered internally consistent by including the integrability condition among their premises.

What do these theories amount to? They are concerned with an individual consumer who is assumed to have a certain amount of money to spend, usually called his income, and to be faced with a fixed set of prices. The satisfaction he gets from what he buys depends solely on the quantities of the different goods and services bought. His action is dictated by choosing that collection of goods and services which maximizes his satisfaction; by the solution, that is, of a constrained maximum problem.

In order to apply this theory it is not necessary to measure satisfaction, or utility, indeed they need not even be measurable. All that is necessary is to work out the implications of the theory which relate to measurable phenomena and check the theory in this way. This amounts to little more than saying that the quantity of each good or service that an individual consumer buys depends on his income and on the structure of prices.

Are we then free to choose any plausible form of relationship relating the quantities bought with income and the price structure ? Not quite, because the theory places some restrictions on the relationships that are strictly compatible with it. In the first place, total expenditure must add up to income. In the second place, there must be no money illusion: the relationships must be homogeneous of degree zero in income and prices. In the third place, the Slutsky condition must be satisfied: the matrix of elasticities of substitution must be symmetric. These considerations apply if we do not assume a form for the utility function, that is the relationship between satisfaction or utility and the quantities of goods and services bought. If we assumed a manageable form for this function we should have determined the form of the demand relationships and these would then turn out to satisfy the restrictions I have mentioned.

I have set up this discussion of demand analysis in terms of a theory in search of observations. The theory relates to an individual consumer. But it is hard to obtain a large number of observations about an individual consumer and even if one could it is doubtful if one would accept the theory as a basis for empirical analysis. According to the theory, the individual's preferences are fixed and so he must not get married or have children, or they will surely change; probably he must not even age to any appreciable extent. His satisfaction depends entirely on his present purchases; it does not matter whether he already has a car or a wireless set. But of course all these things do matter and can be relied on to swamp the influence of income and prices in individual cases.

So before trying to apply the theory, we had better go a stage further and add up the individual demand functions for a whole community of consumers. This will have two effects. First, the particular circumstances of individual consumers will be pooled, and so will largely lose their significance. Second, the income of each individual will appear in the community – or market – function and this will involve an impossible and largely irrelevant number of variables. This difficulty is met by replacing the individual's income by a single average income, to which may be added, if the information is available, the dispersion of the income distribution. In this way we end up with a model of consumers' behaviour in which the consumption per head of each commodity is a function of prices and one or more moments of the income distribution.

We can now apply this theory to observations. The usual way in which this is done is to compile time series of consumption per head of different commodities, prices and average income. The method used is multivariate regression analysis and the technical problems involved are described in detail in chapter XIX of [168]. In carrying out this work it is necessary to modify the model still further by grouping prices. Usually a small number of prices, relating to the commodity we are studying and to a few close substitutes or complements, are included separately and all other prices are combined into a single index-number.

A considerable amount of experience with this kind of model applied to individual commodities is set out in [168]. The main lessons of this experience can be summarized as follows.

1. When applied to a fairly short period of time, such as the inter-war years 1920 to 1938, the model gives an illuminating first approximation to consumers' behaviour.

2. It is usually necessary to recognize that consumers' preferences change over time. An allowance for this is made by introducing a residual trend into the model.

3. The model tends to work better for perishable goods than for durable goods. The reason is that the model is static, so that consumers are always in equilibrium. If circumstances change they will not as a rule be able to adjust immediately to the desired consumption of durable goods because these are usually expensive and pose a problem of finance that cannot be overcome immediately even with the help of hire-purchase. The model can, however, be adapted to meet this difficulty as explained in [164, 165, 166, XIII].

4. In the case of new commodities, the community undergoes a learning process which cannot be represented by a simple, linear or exponential, trend in preferences: a sigmoid trend is necessary. Early writers on this subject, for example Roos and von Szeliski [135] used a symmetrical trend, in fact the logistic. More recently Bain, in his study of television ownership in Britain [10, 11], has shown that a positively skewed trend is to be preferred. He in fact adopts a log-normal integral trend because of its convenient properties [2].

5. If we look only at time series we ignore an important source of information, namely the cross-section data obtained from surveys of family budgets, which might require us to revise our estimates based on time series alone. In fact both sources were used in [168], following the method of extraneous estimators [52].

6. If we consider demand equations in isolation, we run into the difficulty of identification. This is a big subject which I shall consider in the next section.

Thus we see that in trying to apply the classical theory of consumers' behaviour, we are led to a number of major modifications. The important ones relate to changing preferences, including the learning process, and to dynamic considerations connected, in particular, with durable goods. An anti-theorist might reasonably say that the relevant content of the theory was so small and so obvious and the factors omitted by it so important that we could dispense with the theory altogether. Very well, then: let us follow the anti-theorist and start from an empirical standpoint; not, perhaps, from observations in search of a theory but, as we shall see, from observations that may find it hard to get on without one.

Let us suppose that we have time series of all the categories of goods and services bought by consumers, of the prices of these categories and of the number of consumers. We therefore also have time series of separate expenditures per head and of total expenditure per head. Let us now suppose, as a first approximation, that the individual expenditures are linear homogeneous functions of total expenditure and of each price. We may express this assumption by writing, in matrix notation,

$$\hat{p}e = b\mu + Bp \tag{4}$$

where p denotes a vector of prices and $\hat{p}$ denotes a diagonal matrix formed from this vector; e denotes a vector of quantities bought and so $\hat{p}e$ denotes a vector of expenditures; μ denotes total expenditure; and b and B denote respectively a vector and a matrix of parameters.

The model given by (4), although simple, involves a large number of parameters. Thus if consumption is divided into ten commodity groups, there are one hundred and ten parameters to be estimated. This is a formidable number in terms of the length of time series of consumption actually available. So we must try to simplify still further. For this purpose, the natural thing to do is to turn to theory. As we saw above, we should like our demand equations to be additive. The condition that $p'e = \mu$ in all circumstances is satisfied if $i'b = 1$ and $i'B = [0, 0, \ldots, 0]$, where i denotes the unit vector and a prime denotes transposition. We also saw that we should like our demand functions to be homogeneous of degree zero in income and prices. The demand functions in (4) already satisfy this condition and so it adds no further restrictions. The final limitation on individual demand functions that we considered was the Slutsky condition. This is not strictly applicable to market demand functions but except in highly abnormal circumstances we might expect it to hold approximately. If we assume that it does hold and also that the elements of B are independent of p, then, as shown in [148], B must take the form

$$B = (I - bi')\hat{c} \tag{5}$$

where I denotes the unit matrix and $\hat{c}$ denotes a diagonal matrix formed from a vector of constants, c.

If we substitute for B from (5) into (4), we obtain

$$\hat{p}e = b\mu + (I - bi')\hat{c}p$$
$$= \hat{p}c + b(\mu - p'c) \tag{6}$$

In (6) $i'b=1$ and the elements of c are unrestricted. Thus with ten commodities there are nineteen independent parameters to be estimated, a great reduction over the hundred and ten with which we started. This is a manageable number. The properties of this model, its limitations and its advantages, have been set out in [151, 160, 162] and I shall not repeat them here.

I shall however give a simple statement of the consumers' behaviour implicit in (6). The average consumer has a conception of his basic standard of living expressed by the elements of c. He buys these goods at a cost of $p'c$ and so has an amount of money equal to $\mu - p'c$ left over. He spreads this money over the different commodities in proportion to the elements of b.

We may all agree that this is a plausible, approximate model of consumers' behaviour in any particular period. We could hardly suppose, however, that the elements of c and, to a lesser extent, those of b would remain constant over time. It is impossible to add time trends onto (6) without destroying the theoretical properties of the model, but it is possible to make b and c functions of time or any other predetermined variable. Thus writing b_θ and c_θ for b and c at time θ, we could put

$$b_\theta = b^* + \theta b^{**} \tag{7}$$

and

$$c_\theta = c^* + \theta c^{**} \tag{8}$$

When applied to eight broad groups of consumption in Britain over the period 1900 to 1960, the improved model given by (6), (7) and (8) gives fairly good results [160]. In the case of food and clothing these results are in conformity with estimates of total expenditure elasticities derived independently from budget studies. Further refinements of the model and its combination with a budget model are discussed in [153, XIV].

A model is of no practical use unless a means can be found of estimating the parameters it contains. In the present case this can be done [153, 160, XIV]. Also, the model is decomposable and so can be applied hierarchically: we first analyse main groups, then the components of main groups, and so on, until we reach a degree of detail with which the model cannot cope. When we have reached this degree of detail in our practical work we shall have to consider whether other modifications can be derived which will enable us to go further.

We have seen how this particular model can be modified to allow for changes in preferences. It can also be modified to allow for adaptive behaviour. It has an explicit form of utility function [60, 138] which implies ordinal utility and makes possible the construction of a constant-utility index of the cost of living [85]. The whole development could therefore have been given a purely theoretical starting point.

We thus see the development of a model as an iteration between the *a priori* and the empirical. The danger in insisting too much on theories is that they are usually restricted and deal with the phenomena studied on limited assumptions, while at the same time theorists tend to insist on very general formulations within these limitations, formulations which are very hard to manage in practice. The model I have described contains such limitations: it cannot handle complementary or inferior goods; its Marshallian demand curves are never elastic. This would be enough for most theorists; throw it out, they would say. On the other hand, it takes explicit account of changes in preferences and can be modified to represent adaptive behaviour, advantages besides which its limitations within the restricted field of pure demand theory seem, at least to me, rather unimportant.

4. IDENTIFICATION AND SIMULTANEOUS RELATIONSHIPS

I shall now turn to another aspect of economic research which offers scope for both conflict and co-operation between the *a priori* and the empirical. I mentioned in the preceding section that if we consider demand equations in isolation, we run into the difficulty of identification.

The meaning of identification can best be seen from a purely formal example. Suppose that two variables, x and y, are related as follows:

$$y = \beta x + u \tag{9}$$

where β is a parameter and u is a disturbance, and that

$$u = \alpha x + e \tag{10}$$

where α is a parameter and e is a disturbance distributed independently of x. If we substitute for u from (10) into (9), we obtain

$$y = (\alpha + \beta)x + e \tag{11}$$

If we estimate the regression of y on x by the method of least squares we shall obtain an unbiased estimate of $\alpha + \beta$. This will only be an unbiased estimate of β if $\alpha = 0$, which is equivalent to x being distributed independently of u. Thus, given the model consisting of (9) and (10) with $\alpha \neq 0$, we say that β is not identified, meaning that if we try to estimate it by the usual procedure we shall not in fact succeed; what we shall estimate is $\alpha + \beta$. If α is unknown such an estimate will tell us very little about β.

Let us now apply this idea to supply and demand equations. A proposition of elementary economics is that both the quantity demanded and the quantity supplied depend on price. Let us denote series of observations of these three variables by the vectors y_1, y_2 and y_3. Then we can express the simplest possible supply and demand model as follows:

$$y_1 = \alpha_{13} y_3 + u_1 \tag{12}$$

$$y_2 = \alpha_{23} y_3 + u_2 \tag{13}$$

$$y_1 = y_2 \tag{14}$$

where α_{13} and α_{23} denote respectively a demand and a supply parameter, u_1 and u_2 denote series of disturbances and each variable is measured from its mean. If we consider the demand equation, (12), in isolation, the least-squares estimator, α_{13}^* say, of α_{13}, is given by

$$\alpha_{13}^* = y_1' y_3 / y_3' y_3 \tag{15}$$

where, again, the primes denote transposition.

If, now, we premultiply (12) by y_3', divide by the scalar $y_3' y_3$ and take expectations, we obtain

$$\mathcal{E}(y_1' y_3 / y_3' y_3) = \alpha_{13} + \mathcal{E}(u_1' y_3 / y_3' y_3) \tag{16}$$

The left-hand side of (16) is equal to α_{13}^* and can only be equal to α_{13} if $\mathcal{E}(u_1' y_3 / y_3' y_3) = 0$, which it will be if y_3 and u_1 are distributed independently. But, in general, this cannot happen because (14) enables us to equate the right-hand sides of (12) and (13) to give

$$y_3 = (u_1 - u_2)/(\alpha_{23} - \alpha_{13}) \tag{17}$$

so that it would be necessary for u_1 and $u_1 - u_2$ to be distributed independently, a condition that would impose very severe restrictions on the joint distribution of u_1 and u_2.

To make any further progress we need, as usual, more information. Let us consider three possibilities.

First, we may know something about the relative variability of u_1 and u_2. In the extreme case in which $u_1 = \{0, 0, \ldots, 0\}$, the quantity-price relationship in the demand curve, (12), will remain fixed while the corresponding relationship in the supply curve, (13), will shift as a consequence of variations in u_2. Thus what we shall observe will be points on the demand curve and we shall be able to estimate α_{13} but not α_{23}. Similarly in less extreme cases, in which u_1 varies very little compared with u_2, the regression estimate, though biased, will approximate an estimate of α_{13}. It might be thought useful to place some restriction on the joint distribution of u_1 and u_2 and, in more complicated examples, this would probably be the case.

Second, further reflection might suggest a modification of the original model. In (13) it might seem reasonable to assume that this year's supply depended on last year's price because of the time lag, as in agriculture, between the decision to produce and the emergence of the product. Thus in (13), but not in (12), we should replace y_3 by its value in the preceding year. With this change, α_{23} can now be estimated consistently. Also, as a consequence of (14), the right-hand sides of (12) and the new version of (13) can be equated to express y_3 in terms of its value in the preceding year. The coefficient α_{23}/α_{13} can thus also be estimated consistently and its reciprocal multiplied by the estimate of α_{23} will provide a consistent estimate of α_{13}. This is a very simple example of what Bentzel and Wold have called a recursive system [18]. In such a system the simultaneity of causes gives way to a sequence of causes, and the equations can be solved one after the other in a definite order. In my example, last year's price determines this year's supply which in turn, together with the market-clearing assumption, (14), determines this year's price, apart from a disturbance of the form $(u_2 - u_1)/\alpha_{13}$, and at the same time this year's demand, apart from a disturbance, u_2.

Third, it may be possible to extract from the disturbances and introduce explicitly some of the other demand and supply variables, such as consumers' income, x_1 say, and rainfall, x_2 say. We could then rewrite the model as

$$y_1 = \alpha_{13}y_3 + \beta_{11}x_1 + u_1 \tag{18}$$

$$y_2 = \alpha_{23}y_3 + \beta_{22}x_2 + u_2 \tag{19}$$

$$y_1 = y_2 \tag{20}$$

where it is assumed that x_1 and x_2 are distributed independently of u_1 and u_2. No new symbols are given to the disturbances, but it will be recognized that the new disturbances are different from the old ones in (12) and (13).

If, using (20), we replace y_2 by y_1 and if we take the terms in y_3 to the left-hand side, we can rewrite the new model in the form

$$[y_1 y_3]\begin{bmatrix} 1 & 1 \\ -\alpha_{13} & -\alpha_{23} \end{bmatrix} = [x_1 x_2]\begin{bmatrix} \beta_{11} & 0 \\ 0 & \beta_{22} \end{bmatrix} + [u_1 u_2] \tag{21}$$

or, more compactly, as

$$YA = XB + U \tag{22}$$

whence, on postmultiplying by A^{-1}, we obtain

$$[y_1 y_3] = \frac{1}{\alpha_{13} - \alpha_{23}}\left\{[x_1 x_2]\begin{bmatrix} -\alpha_{23}\beta_{11} & -\beta_{11} \\ \alpha_{13}\beta_{22} & \beta_{22} \end{bmatrix} + [u_1 u_2]\begin{bmatrix} -\alpha_{23} & -1 \\ \alpha_{13} & 1 \end{bmatrix}\right\} \tag{23}$$

or, more compactly,

$$Y = XBA^{-1} + UA^{-1}$$

$$= XC + UA^{-1} \tag{24}$$

where $C \equiv BA^{-1}$.

On the assumptions made, each equation in (23) can be treated separately by the method of least squares to give a column of the regression estimator, C^* say. From C^* we can obtain consistent estimators of the original regression coefficients. Thus $\alpha_{13}^* = c_{21}^*/c_{22}^*$; $\alpha_{23}^* = c_{11}^*/c_{12}^*$; $\beta_{11}^* = (c_{11}^* c_{22}^* - c_{12}^* c_{21}^*)/c_{22}^*$; and $\beta_{22}^* = (c_{12}^* c_{21}^* - c_{11}^* c_{22}^*)/c_{12}^*$.

This example has been chosen so that the equation $BA^{-1} \equiv C$ has a unique solution from which A and B can be determined. In this case the system of equations (22) is said to be identified. It may happen that the solution is not unique, in which case (22) is said to be over-identified and a more elaborate procedure is required to extract a set of best estimates from the discordant solutions [44, 89].

The problem that has just been illustrated arises simply from carrying into the field of statistical estimation the multiple dependencies between economic variables which have been recognized by theorists for the last hundred years. Economists were surprised when in 1914 Moore published an empirical demand curve for pig-iron in which the quantity demanded varied positively with the price of pig-iron [109]. In the 1920's, the essence of the difficulty was laid bare by Working [191] and later the statistical problem was solved, on the above lines, by Haavelmo [67]. The whole question was studied in great detail at the Cowles Commission [44, 89]. More recently an elegant account of its wider importance has been given by Hurwicz [73] and the statistical problem has been revisited by Durbin [53].

The solution illustrated in this section applies to a real problem in building economic models and is, theoretically, incontrovertible. In applied work, however, we must, as always, weigh the advantages to be gained from it against the costs involved, the money or time which could be used to handle other difficulties which may seem even more pressing. In trying to reach a decision, we should take the following considerations into account.

First, the construction of a system of equations round a single equation can only be expected to improve the estimates of the regression coefficients by bringing into account in a realistic manner some phenomena which are closely connected with the single equation. A formal system, built to satisfy formal conditions of identifiability, is not likely to be helpful. Before committing ourselves to this work, however, we should satisfy ourselves that simpler methods will not serve our purpose. For example, Koopmans has shown in [86] that, in the limiting case of perfectly elastic supply, we can use a single demand equation with the amount demanded as the dependent variable and that, in the case of perfectly inelastic supply, we can use a single demand equation with the price as the dependent variable. Also, as I have said, if we can reduce our model to a recursive system we can properly make use of single equations.

Second, even if we are not in one of the fortunate positions just considered, we may reasonably ask whether the bias in simple least-square estimates which could be removed by the use of simultaneous equations is likely to be large. Unfortunately it is difficult to determine even the order of magnitude of the bias in different situations, though a certain amount of empirical information is accumulating [19, 55].

Third, the consistency at which simultaneous equation methods are directed will not be achieved if the predetermined variables are random variables measured subject to error. For it can be shown that such variables are not, in general, distributed independently of the disturbances. Thus the estimates which enter into our calculations, as opposed to their true values, cannot properly be regarded as predetermined.

Finally, the determining variables in the transformed equations, such as (24) above, must be distributed independently of the disturbances in those equations, and these depend in turn on the disturbances in many, if not all, of the original equations. In practice this condition is likely to be quite severe.

The conclusion of this discussion is open. Those who are wedded to simultaneous equation methods will help to build up the experience we need, but those who are not may well be right in believing that their research effort can be better expended in other directions.

5. INFINITE HORIZONS AND THE PROBLEM OF PLANNING

The third and last problem I shall consider in this paper has to do with economic planning. The problem is a restricted one but, I believe, of considerable importance in formulating plans for the future. It can be stated as follows.

What we do this year influences what we shall be able to do next year, and what we do next year influences what we shall be able to do the year after, and so on indefinitely. Thus from a theoretical point of view we are led to contemplate an infinite time-horizon, since there is no reason to stop at any particular point. In other words there is no date in the future of which we can say *après ça le déluge* and at which we can plan for productive capacity to come to an end.

On the other hand, we can know very little about the remote future and it does not seem reasonable that what we do today should be influenced by the state of the world hundreds, to say nothing of millions, of years from now. We may try to diminish the importance of the future in our calculations by applying a rate of time-discount. This will certainly have its effect, but it is a double-edged weapon. If our rate of time-discount were x per cent we should have no reason to do an investment that did not yield x per cent, and if x were high enough we might even be led to quite a lot of disinvestment. But this cannot be our intention. No, we had better face the problem of infinite time-horizons squarely. If we do, we must specify not only what is to happen over a finite future period but also what is to be available at the end of it with which to carry on economic life.

Thus we see that an *a priori* approach would enable us to solve the problem of the infinite time-horizon but would require of us an unbounded knowledge. Since we do not possess this knowledge we must adopt an empirical approach. Let us now illustrate the problem from these two different points of view.

The *a priori* approach can be illustrated by Ramsey's theory of saving [131]. The problem is to determine the optimum rate of saving now and the method is to consider that rate which will maximize the total utility to which the economy gives rise, considered over an indefinitely prolonged future.

The problem and its theoretical solution can be set out very simply for an economy that produces a single good which can be either consumed or added to the capital stock. This is done at the beginning of section 4 of [V] on p. 59 below. As can be seen there, three functions are required: a function, f, which relates output to inputs of labour and capital; a function, ϕ, which relates the utility of consumption to the amount of the good consumed; and a function, ψ, which relates the disutility of labour to the amount of work performed. The solution is given in (27) on p. 61 below, from which it appears that saving should equal the difference between bliss (the maximum utility to which the economy can give rise) and the present level of utility, divided by the marginal utility of present consumption. If for the sake of argument we give simple, manageable forms to the functions f, ϕ and ψ, we can reach explicit solutions for the variables of the model in terms of the parameters of these functions. An example of this is given, again, in [V] (pp. 61–4).

This theoretical model certainly gives great insight into the nature of the ideal saving behaviour and correctly emphasizes the variational character of planning problems. The difficulty with it from the practical point of view is that we have to specify the functions f, ϕ and ψ so that they hold good for all time. This we clearly cannot do. So, as usual, we must turn to a manageable model, preserving as far as possible the theoretical insights we have gained.

One such model is being developed by a group of us at Cambridge in connection with our study of the possible growth of the British economy. It would take me too long to set down the model in algebra but I will try to give a verbal summary of it.

The model consists of two parts, a short-run model [VIII] and a long-run model [30, 159]. The first of these deals with a transitional period, say from now to 1970, and the second deals with developments after that date. If we think of the growth rates of the outputs of different products as the elements of a vector, the purpose of the long-run model is to choose a length and direction for this vector, bearing in mind the rate at which we should like consumption to grow, the changing composition of consumption, the labour available, the balance of payments, technical developments and so on. To expand in this way from 1970 on means that we must have an initial stock of assets in the different industries which will enable the available labour to produce the required output. The initial stock of the long-run model is the same as the terminal stock of the short-run model whose purpose, accordingly, is to maximize consumption over the transitional period subject to certain constraints including the need for a minimum vector of terminal stocks of different kinds of assets. Thus, in the simplest case, the initial stock of the long-run model is determined by the desired level of consumption in 1970 and the rates of growth of its components, while the same initial stock determines the time paths of the components of consumption over the transitional period. We can now examine the total time paths of consumption and its components. If they form acceptably smooth series from the past, through the transitional period and into the long-run state of steady growth, we can accept the solution as a policy. If they do not, we must reconsider the initial conditions of the long-run model. For example, we should almost certainly improve things if we reduced the level of consumption from which we started the long-run period of steady growth. Eventually, a policy would emerge from an iteration between the two models. This approach has a good deal in common with Ramsey's. The short-run model involves the solution of a variational problem, and this can be done by the method of dynamic programming [16, 17]. The difference between our model and Ramsey's is that in place of the infinite time horizon we have a desired direction and rate of growth relating to the comparatively near future. This date should be fixed so as to allow time for adaptation which depends partly on the period of production of large capital goods and partly on human adaptability in such matters as mobility, education and training, attitudes to innovation and capacity for administrative change.

Thus, we see again the interplay of the *a priori* and the empirical.

6. CONCLUSION

I have only one conclusion to draw from this survey of economic problems: the *a priori* and the empirical are completely intertwined in the development of the subject. True, there are specialists whose interests are mainly empirical or mainly theoretical. The first work with a rather small input of theory, the second with a rather small input of facts. But this is not the normal, healthy case; indeed, modern developments in economics have come about largely through breaking down barriers between excessive specialisms. I do not think that most of us could separate the contributions of the *a priori* and the empirical in the development of our ideas nor do I think that most of us would wish to deny the essential part played by each.

III

THE HOUSEKEEPER AND THE STEERSMAN

M. Jourdain: Quoi? quand je dis: 'Nicole, apportez-moi mes pantoufles, et me donnez mon bonnet de nuit', c'est de la prose?

Maître de Philosophie: Oui, monsieur.

M. Jourdain: Par ma foi! il y a plus de quarante ans que je dis de la prose sans que j'en susse rien.

J. B. MOLIÈRE

1. SUBCONSCIOUS CYBERNETICS

M. Jourdain's attitude to prose characterizes the attitude of many economists of the present day to cybernetics; for years they have been talking cybernetics without knowing it.

Cybernetics may be described as the study of communication and control in self-regulating systems. It is perhaps of interest to realize that an economy can be regarded as such a system and that some economists have in fact regarded it in this way for the last hundred years. On this view, an economy is designed to meet human needs for consumption and this involves the production of goods and services and the accumulation of durable equipment needed in the process of production. This system is regulated on the principle of self interest by human beings organized in markets and responding to price signals. When television is invented it tends, after a period in which consumers learn its potentialities, to replace other forms of entertainment. Sales increase, the price falls with the increase in the scale of production and with technical developments, sales are further stimulated, and a new equilibrium tends to be established with more television sets and fewer cinemas. A change in tastes, a change in technology, the failure of a harvest, the repeal of a law, the outbreak of a war, alter the circumstances in which this endless tendency to adjustment is played out, so that the human agents in the economy, like a vast human computer set off on an endless tracking project, grope their way towards the moving equilibrium which is their implicit goal.

In his famous *Eléments* (1874), Léon Walras wrote: 'Tel est le marché permanent, tendant toujours à l'équilibre sans y arriver jamais par la raison qu'il ne s'y achemine que par tâtonnements et qu'avant même que ces tâtonnements soient achevés, ils sont à recommencer sur nouveaux frais, toutes les données du problème...ayant changé' [186, pp. 369–70]. As

Goodwin [65] has pointed out, Walras sometimes speaks as if his presentation of economic adaptation were purely formal and sometimes as if it were a description of the way in which adaptation takes place in a free-enterprise economy.

The Italian economist Barone, writing early in this century [12], took another step in this interpretation of the economic process. He perceived that the idealized workings of a free-enterprise economy could be reduced to rules and that if an all-powerful Ministry of Production could require all economic agents to observe these rules, a system would emerge which, in the interests of efficient planning, would reproduce the workings of an idealized free-enterprise economy. Barone was concerned to show that little could be achieved by planning which could not be achieved without it. More recently, writers such as Lange [91] and Lerner [96] have used essentially the same arguments to show that, by an appropriate setting of the rules, a socialist state could make an economy work in the idealized way, very different from reality, which many writers have claimed for the free-enterprise economy.

In 1931, Kahn published his famous paper [78] on 'the multiplier', a term which relates to the closed loop connecting income and consumption. If an extra £1 is spent on goods and services, total income will rise by £1. If a proportion, k, of an increase in income is spent on goods and services, the extra expenditure of £k will lead to extra income of £k and so to still further expenditure of £k^2, and so on indefinitely. Thus the original spending of £1 will lead to induced expenditure of £$(k+k^2+...)$ and the total additional expenditure and income will be £$1/(1-k)$. The sum $1/(1-k)$ is called the multiplier: if $k=\frac{1}{2}$, the multiplier is 2; if $k=\frac{2}{3}$, the multiplier is 3 and so on. This idea plays an important role in Keynes' *General Theory* [82] and generalizes readily from the income and expenditure of a single sector to the income and expenditure of any number of interdependent sectors. Thus if A denotes a convergent matrix whose elements, a_{rs}, measure the proportion of the income of sector s spent on the products of sector r, then the expression $(I-A)^{-1}$, where I is the unit matrix, is termed a matrix multiplier. This generalization is at the centre of input–output analysis, the name given by its originator Leontief [93, 94] to the scheme now generally in use for analysing industrial interdependence.

It is recognized by most people that actual economic systems do not work perfectly if left to themselves: in particular the level of production tends to fluctuate. This raises the question of whether it is possible to introduce automatic stabilizers which would reduce the fluctuations to within pre-assigned limits. Thus in 1938 Meade [106] proposed a scheme of consumer credits and taxes on employment (negative credits), varying automatically with an index of unemployment, as a means of building stability into the economy. These ideas were reflected in the White Paper on employment policy [179] issued by the British Coalition Government in 1944.

These are a handful of examples out of a very large number which show economists viewing the economy as a self-regulating system, with particular problems of communication and control, long before cybernetics became a popular term. I have called this section 'subconscious cybernetics' not in any disparagement of the outstanding writers I have mentioned but because they were mainly concerned with a theory of the economic process expressed in words or in very general mathematical terms, and did not either specify the closed-loop system they were describing or, with the exception of Leontief, attempt to measure its operating characteristics. In fairness it should be said that before the systematic collection of economic statistics on a large scale, a development that was greatly stimulated by the first world war and even more by the second, such measurement would have been impossible; and that without the invention of modern computers it would not have been possible to reach solutions with any but the most highly aggregated models.

2. CONSCIOUS CYBERNETICS

Since the last war, a number of writers have looked at economic models, or parts of economic models, from the point of view of control-system engineering, and closed-loop diagrams have become a feature, though hardly a central feature, of economic literature. In this section I shall give a number of examples.

(a) *Equilibrium in isolated markets.* In the paper by Goodwin already referred to [65] the market adjustment schemes of Walras and of Marshall are discussed and illustrated by means of a closed-loop diagram. In the Walrasian scheme, the amount demanded, q_d, is a function $D(p)$ of the price p, and the amount supplied, q_s, is a function $S(p)$ of the price. If we start with an arbitrary price we can consider the conditions under which the market will eventually attain equilibrium. A solution, that is an equilibrium price, requires that $D(p) - S(p) = 0$. If we add a first value of p to the left-hand side of this equation we can calculate a second value; if we then add this to the left-hand side we can calculate a third value, and so on. This process will converge provided that the slope of the demand curve, $D(p)$, is greater in absolute value than the slope of supply curve, $S(p)$. This process is illustrated by Goodwin in a diagram which, for comparison with other diagrams in this paper, can be drawn as shown in diagram 1.

In this diagram, the variables are enclosed in circles and the relationships are shown along branches. The operator E denotes the next value in an ordered sequence. Given an initial price, p, this determines Eq_d through the demand function $D(p)$ and Eq_s through the supply function $S(p)$. The next approximation, Ep, to the equilibrium price is given by the relationship $Ep = Eq_d - Eq_s + p$. By multiplying Ep by E^{-1} we can regard it as the new initial price and continue in this way until convergence produces a solution.

4

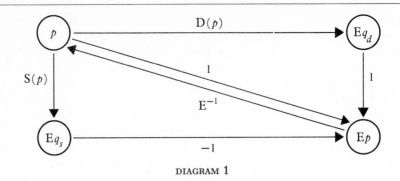

DIAGRAM 1

In this model of the market adjustment process the price is successively changed until equality is obtained between supply and demand. Alternatively, following Marshall, we can consider a process in which the quantity is successively changed until equality is obtained between the demand price and the supply price. This model can be represented on diagram 1 if p and q are interchanged and $D(p)$ and $S(p)$ are replaced by their inverse functions.

(b) *Macro-dynamic models.* The main users of closed-loop diagrams to express economic relationships are engineers or writers with engineering experience. For example Tustin has made a number of interesting contributions to economics [172, 173] in which a variety of simple models are set out in this way.

We can illustrate this work by considering the familiar multiplier–accelerator model. This model has two loops connecting income to itself. In the first, the connection is made *via* consumption using the multiplier; in the second it is made *via* capital expenditure or investment using the accelerator, that is a relationship connecting investment with the rate of change in income. The model is shown in finite difference form with simple time-lags in diagram 2.

In this diagram A denotes autonomous expenditure, Y denotes income, C denotes consumption and V denotes investment. An increase in autonomous expenditure leads to an equal increase in income and this effect is then magnified in two ways. First, there is the consumption sequence where it is assumed that consumption in this period is proportional to income in the preceding period, so that the multiplier is lagged. Second, there is the investment sequence where it is assumed that investment in this period is proportional to the change in income between the preceding period and the period before that, so that the accelerator is also lagged. Many variants of this simple model can be found in the literature. An early and particularly interesting one in which the details of the investment loop were elaborated, the stock of assets and the investment decision process being introduced

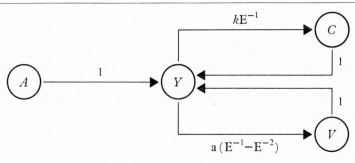

DIAGRAM 2

explicitly, was published before the war by Kalecki [79, 80, 81]. Other examples are due to Goodwin [66] and Phillips [120]. Many of these models are brought together and set out in uniform diagrams by Allen [5, 6].

(*c*) *Economic control.* Phillips, another writer with engineering experience, has applied the methods of control-system engineering to the problem of economic stabilization [120, 121, 122]. The simplest case of a multiplier model with error correction, which will now be illustrated, links up with Meade's work on consumer credits but clarifies the knowledge required for an efficient control scheme.

Reduced to its simplest terms, Phillips' scheme is shown in diagram 3.

As before, A and Y denote respectively autonomous expenditure and income (or output). Two other variables now appear: X, which denotes expenditure and Y^* which denotes the level of income aimed at. The symbol μ denotes a distributed lag and k, as before, is the proportion of income which is spent. The stabilizer is shown as $\lambda F(\epsilon)$, that is a function of the error, $\epsilon = Y^* - Y$, which is fed back to expenditure subject to the distributed lag, λ. For efficient control this function should contain three

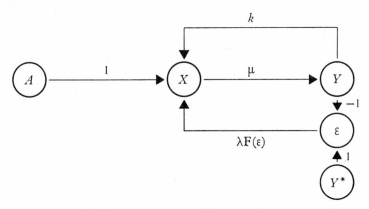

DIAGRAM 3

components depending respectively on the current value of the error, its integral over the past and its rate of change. In terms of Meade's example this means that the consumer credits released or the employment tax imposed should be related to the persistence of unemployment and to its rate of change as well as to its level. Exactly how it should be related to these factors depends on the responses in the system and the lags and time delays to which these are subject.

Phillips recognizes that, even in principle, the problem of economic stabilization is extremely intricate and that a detailed knowledge of economic relationships is needed before policy recommendations can be made. Nevertheless, he is able to draw a number of conclusions about the desirable characteristics of economic stabilizers.

3. CYBERNETICS IN ECONOMIC ANALYSIS AND POLICY

The illustrations of the last section, with the exception of Phillips' work on control, represent the translation of economic theories into a different language from the one in which they were originally expressed. By doing this we may hope to gain several advantages: we can consider the use of analytical techniques which are certainly not to be found in the economist's normal tool kit; we can often see more clearly the similarities and differences between rival theories; and we can hope, by trying to talk the language of engineers, to spread an understanding of economic problems to a group who are potentially of great importance in the development of realistic economic models.

I shall now turn to a number of fields of cybernetic interest in applied economic research and comment briefly on these.

(a) *Economic planning.* About eighteen months ago I described [152] a model of the growth of the British economy on which a group of us here is working, and a fuller account of it has now been published in [30]. This is a large, computable model dealing with production, consumption, accumulation and foreign trade. The basic data on economic flows are arranged in a social accounting matrix containing 253 accounts, the relationships are based mainly on an analysis of past statistical information and are assumed to change in a measurable, systematic way, and the whole model is programmed for the Cambridge computer, EDSAC 2. The principal purpose of the model is to show how economic balance could be achieved on different assumptions about the rate of growth of total consumption, abstracting in the first instance from transitional problems of adjustment. The object of this exercise is to give a high but manageable target for growth. If this could be decided upon it would be worth while to work on the transitional problems.

Such a model can naturally be regarded as a multiple-loop feed-back system. For example a target must provide for a reasonable degree of balance between imports and exports, and so the model must contain a means of bringing this balance about. This is essentially an iterative problem because one does not know what one will need to import until one knows what one is going to produce and one does not know what one will have to produce until one knows what one can afford to import. Thus one feed-back loop is concerned with foreign trade.

Another is concerned with industrial capital expenditure. If the demand for goods and services is to grow at preassigned rates, the different branches of industry producing these goods must expand their capacity appropriately. This means capital expenditure; but how much and where? If we consider different rates of exponential growth in the components of consumption, the answers depend on summing an infinite series of vectors the elements of which fortunately damp out fairly quickly.

(b) *Shadow prices*. Another important loop concerns the future relative prices of the different products and the effect these may be expected to have on the composition of consumption. In trying to calculate future relative prices we have to take account of technical progress and of the allocation of labour and capital to different industries. We have also to take account of the probable responses of consumers to changes in relative prices so that we are again involved in a convergent iterative process.

In principle, these ideas could be applied to the balance of payments problem, but for the time being we have to fall back on cruder means of obtaining a balance because to allow fully for price adjustments we should need the future shadow prices of other countries as well as of our own. If we had this information we could not only determine foreign trade by reference to the usual price signals but we should also find that equilibrium exchange rates would emerge as the elements of a characteristic vector of an international trading matrix. This may sound unspeakably 'brave new world' but the countries of the European Economic Community are already beginning to build a set of integrated multi-sector models for their economies, so it may not be as far off as many people would suppose.

(c) *Computers*. Quantitative models of the size I am describing cannot be analysed without a modern computer. Thus our tool itself encourages us to use iterative processes. In the present case we need a programme which does not have to be rewritten every time we want to change one or other of the relationships of the model. To this end we have programmed the model in stages and have chosen computing algorithms which are not dependent on linearity.

(d) *The human element*. It is a mistake to suppose that realistic economic models can be constructed, let alone applied, without a sympathetic understanding on the part of many people who are not economists. In the first place, the technical and behavioural relationships represented in economic models are always on the move and a direct attack on these problems is to be preferred to the indirect attack which is often the best that economists can manage on their own. This means interesting technicians and business economists in general economic models to the point of inducing them to put their special knowledge at the disposal of the model builders. In the second place, major changes of economic behaviour involve changes in individual attitudes and accepted ways of doing things. This means convincing a much wider public from whom a sympathetic hearing may be harder to obtain.

4. CONCLUSION

The main theme of this note is that economics and cybernetics are closely related because the search for equilibrium is characteristic of economic systems. Indeed our ability to keep the economy on a chosen course has become nowadays a central political issue. Economics therefore shares a cybernetic aspect with other fields of scientific research as different as ecology, neuro-physiology and engineering. Conversely, many sciences which have nothing to do with human societies have an economic aspect in that the objects of their analysis, whether they are spider's webs, honeycombs, soap bubbles or brachistochrones, illustrate in their different ways nature's solution to the problem of the efficient use of resources.

IV

MISERY AND BLISS

1. INTRODUCTION

A word of explanation is needed at the outset to indicate the way in which this paper, which was presented at the World Population Conference (Rome, 1954), fits into the sessions on 'Population in relation to Capital Formation, Investment and Employment'. The author was originally asked to write a short paper on *Relations between Consumption Patterns and Rate of Savings and Population Growth in Underdeveloped Countries*, it being recognized that a general lack of basic knowledge in this field would probably make it desirable to define the relationships studied and the problems to which they give rise before it would be possible to discuss the effects on them of population growth.

After a preliminary survey it seemed to the author that it would be useful to try to map out a theoretical framework in terms of which the subject and its policy implications could be discussed. There exists of course the classical population theory of Malthus in which consumption behaviour plays an important role. In recent years a large amount of work has been done on the determinants of spending and saving and also on models of economic growth which embody a consumption or saving function and which can be formalized in terms of a multiplier–accelerator model. Something more however is needed. Spending and saving patterns can lead to difficulties even if the Malthusian connection between population growth and the standard of living does not operate. As for more recent models their is, first, the problem that from an empirical point of view the acceleration principle does not provide an adequate formulation of the relationship of real asset formation to output, and second, that models of this type do not readily permit a contrast between what is actually likely to happen when a steady state is reached with what might be expected under ideal conditions. It may be asked why the two states, the actual and the ideal, should be different. The answer must lie, apart from random and purposeless behaviour, which of course may exist in practice, in the fact that the aims which govern the spending and saving behaviour of individuals may not be compatible with an optimal exploitation of the community's technological environment.

This situation provides the point of departure for the present paper. Its object is to examine some of the economic problems involved in population growth and adaptation. The model of actual behaviour (positive theory) set out in section 2 is simple and, as a consequence, highly abstract. A central element of it is a production function which expresses the fixed technological environment that forms part of the circumstances of life of the hypothetical community studied. If the paper were concerned primarily with rich communities, accustomed to technological change and possessing a considerable degree of adaptive behaviour it might not be very illuminating to summarize the conditions of production in this way. For such communities may be supposed usually to be fairly well adapted to their technological environment as it is and their main concern is with the development of the largely unknown technological environment of the future. In the poorer communities of the world today on the other hand, and especially in those which have seen little economic or demographic change for long periods in the past, a production function which summarizes the position as it is or as it very well might be from both technological and sociological points of view may be an illuminating concept, since an important form of adaptive behaviour required of such communities is that involved in moving from a given production situation to one which takes advantage of the many technological advances that have taken place in Western economies and may in consequence be expected to result in a higher standard of living.

But even if the setting of the problem is promising, any theory and particularly a rather simple and all-embracing theory like the present one depends on the characteristics of the various relationships of which it is composed. If, therefore, the theory is thought to possess some plausibility and interest it is desirable to discover the extent to which the various relationships are in fact capable of describing observed behaviour or technology and also to form a view at least of the orders of magnitude of the parameters that enter into them. No work of this kind is contained in the present paper, and when actual behaviour is mentioned, what is meant is the behaviour which emerges from the positive theory.

The positive theory is followed in section 3 by a normative theory. In this part one of the relationships of the positive theory is replaced by a relationship derived from the requirement that assets be maintained at a level which will make the standard of living that can be maintained as high as possible. A comparison of the two theories provides some indication of the circumstances in which actual behaviour will prove ill-adapted to the policy aim considered.

The possible discrepancy between the actual and the optimal standard of living leads to the necessity of a policy prescription designed to carry the community from the actual to the optimal state. In section 4 such a prescription is introduced in the form of a saving principle which indicates how much a community ought to save if it is to attain the highest standard of living which its technology will allow, and how that saving should be distributed over time in order to yield the largest possible stream of consumption. Such a principle was introduced rather more than a quarter of a century ago by F. P. Ramsey [131] and it appears to provide precisely what is needed for the present purpose. Ramsey's treatment is restricted to the case of a stationary population but it can be adapted to the case of a growing population as is done below.

It is evident from this brief outline that the paper is arranged under general headings appropriate to any policy subject and is concerned with what would actually happen on a given set of hypotheses, what ought to happen and the means of passing from the first state to the second in the case of a difference between them. This arrangement which seems appropriate to the subject matter is intended to provide an analytical study of a certain type of problem conducted on highly simplified assumptions. It does not purport to provide practical answers to practical questions. For that more difficult task it would be necessary to consider the relevance of the model to any particular case and to extend it if necessary so that social and political considerations and also other economic influences could be given the weight due to them.

The title of the paper reflects both the condition of a community governed by Malthus's classical theory of population and the objective of policy as stated by Ramsey in his theory of saving. 'Misery' was the principal means whereby the populations of the Malthusian theory were kept in check, but it also provides a good description of their condition, and it is quite possible for a relatively miserable state of affairs to persist in the absence of sufficiently strong adaptive behaviour even if Malthus's principle does not drive a population to the level of subsistence. 'Bliss', on the other hand, is the convenient term used by Ramsey for the situation reached when full advantage is taken of the potentialities of available technology.

In concluding this introduction I should like to acknowledge the help I received in preparing this paper from my friends Dr S. J. Prais and the late Dr R. E. Brumberg.

2. POSITIVE THEORY

This section is devoted to the elaboration of a simple theory of economic growth and the maintainable states associated with it. The essential variables considered are the rate of population change, which at certain stages in the argument is treated as a parameter, and various accounting magnitudes such as the stock of assets and the flow of consumption. The accounting variables are connected by a certain number of accounting identities. In addition four other relationships are introduced. The first is a production function which sets out the possibilities permitted by the fixed technological environment of the economy. The second indicates the level of asset formation required to maintain a given level of assets per head. The third indicates the determinants of spending or saving. The fourth relates the rate of population growth to the standard of living achieved by the economy. The first three of these relationships taken together with the accounting identities are used to determine the maintainable values of the accounting variables for given values of the rate of population growth. If the final relationship is introduced the rate of population growth is also determined and it can no longer be treated as a parameter.

(*a*) *The accounting identities.* The accounting identities are those which arise in the consolidated national accounts and balance sheet for a simple closed economy. Variables denoted by small letters relate to quantities per head of the population while those denoted by capitals represent aggregate quantities.

The first identity states that purchases equal sales on the consolidated production account, that is

$$y \equiv c+i \tag{1}$$

where y denotes income (or output) per head, c denotes consumption per head and i denotes asset formation per head.

The second identity states that receipts into the consolidated capital transactions account equal the outgoings from that account, that is

$$s \equiv i \tag{2}$$

where s denotes the flow of saving per head.

The third identity connects the opening and closing stocks of real assets in the balance sheet, that is

$$EA \equiv A+I \tag{3}$$

where A denotes the stock of real assets at the beginning of the period and E is an operator which advances by one time unit the variable to which it is applied. Thus (3) defines the closing stock of assets in the balance sheet as the opening stock plus the asset formation of the period.

(*b*) *The production function.* The economy considered makes use in production of two factors, labour and real assets. Its output is conceived of as a single commodity, which may be used either for consumption or for making additions to the stock of assets. The economy is technologically stationary in the sense that output depends simply on the amounts available of the two factors and is not shifted upwards or downwards in relation to these amounts by any change in technical knowledge or other influences. An essential feature is that there is an optimum ratio of assets to manpower at which output per head is a maximum. Thus it may not be possible for the total demand for output to be met by one man if only he has sufficient assets to work with.

A production function with these characteristics could take many forms. A simple and convenient one which will be adopted here is

$$y = y^* - \beta(a-a^*)^2 \tag{4}$$

where a denotes the stock of real assets per head. The symbols y^*, a^* and β denote parameters of the production function. In particular y^* denotes the highest attainable output per head which is reached when the actual amount of assets per head, a, is equal to a^*. If Y, A and n denote respectively total output, total assets and total manpower, then (4) can be written in the form

$$Y = ny^* - \beta\left(\frac{A}{\sqrt{n}} - a^*\sqrt{n}\right)^2 \tag{5}$$

whence, as is otherwise obvious from Euler's theorem on homogeneous functions, it can be confirmed that

$$Y = \frac{\partial Y}{\partial n} \cdot n + \frac{\partial Y}{\partial A} \cdot A \tag{6}$$

so that, if each factor is paid its marginal product, the total product will be exactly distributed without residue.

A feature of this production function is that saving is at best useful to attain or maintain the optimum assets–manpower ratio, a^*. Beyond this point it would certainly be useless to urge the importance of thrift on the members of a community the technological environment of which was characterized by (4). Greater thrift could in the most favourable circumstances only be useful (and not positively harmful) to such a community if it were accompanied by technological change which would permit the use of more capital per head. Such a situation can be described in a different way as one in which capital, though perhaps scarce by comparison with accepted usages elsewhere, may be a free good in a community with only a limited knowledge of how to use it. Beyond a certain point such knowledge is a prerequisite of self-help by means of greater thrift.

(c) The maintenance of assets per head with steady growth of population.
If a population is in a state of steady growth with constant age composition and if it grows in one time period from 1 to λ then it is possible to derive from (3) the asset formation per head needed to maintain a given level of assets per head. For (3) can be written in the form

$$I = (E-1)A$$
$$= a(E-1)n$$
$$= a(\lambda-1)n \tag{7}$$

whence the relationship

$$i = a(\lambda-1) \tag{8}$$

obtained by dividing both sides of (7) by n shows the asset formation per head, i, required to maintain assets per head constant at the level a.

(*d*) *The saving function.* It is proposed to adopt here an aggregate saving function of the form

$$S = (1-\gamma)\,Y - \delta A \tag{9}$$

where γ is the marginal propensity to spend out of income for given assets and δ is the marginal propensity to spend out of assets for given income. Equation (9) represents an obvious generalization of the more usual function in which saving is assumed to be proportional to income and indeed reduces to that form if $\delta = 0$. It has been shown by Modigliani and Brumberg [107] that (9) can be derived from an individual saving principle in which each individual saves solely to provide for his own retirement and that a relationship of this form can be used to explain certain of the known facts about aggregate spending and saving in the United States. While the economies to which the models considered in this paper may be supposed to apply differ in many ways from that of the United States, in the absence of any other detailed investigations it would seem useful to consider the implications of the saving behaviour implied by (9).

(*e*) *The influence of the standard of living on population growth.* It is here supposed that λ depends on c, the standard of living, and, specifically, that

$$\lambda = \theta + \mu c \tag{10}$$

Following Malthus it would be necessary to suppose that μ is positive but (10) can also accommodate an anti-Malthusian type of reaction in which a higher standard of living is accompanied by a lower rate of population growth.

(*f*) *A summary and interpretation of the positive theory.* If (1) and (2) are used to eliminate i and s it can be seen that the four variables, a, y, c and λ are connected by the following four equations

$$\left.\begin{array}{l} y = y^* - \beta(a-a^*)^2 \\ y = c + a(\lambda - 1) \\ c = \gamma y + \delta a \\ \lambda = \theta + \mu c \end{array}\right\} \tag{11}$$

If a substitution for c is made from the third into the second equation of (11) there results

$$y = \frac{(\lambda + \delta - 1)}{(1-\gamma)}\,a \tag{12}$$

If λ is treated as a parameter the maintainable steady states can be shown diagrammatically by drawing the relationships between y and a as given by (4) (the first equation in (11)) and (12). Equation (4) will give a parabolic relationship while (12) will give a set of straight lines through the origin, one relating to each value of λ, such that the larger the value of λ the steeper the slope of the line. The position is illustrated in diagram 1 in which the assets–output ratio for maximum output is assumed to be $4\frac{1}{2}$ to 1 and the parameters of the saving function are based on the researches of Modigliani and Brumberg.

For different values of λ, the maintainable steady states are given by the intersection of the appropriate line and the parabola. This means that a community whose behaviour is characterized by the first three equations of (11) will eventually reach a steady state in which the values of a and y are given by the coordinates of the point of intersection just referred to and in

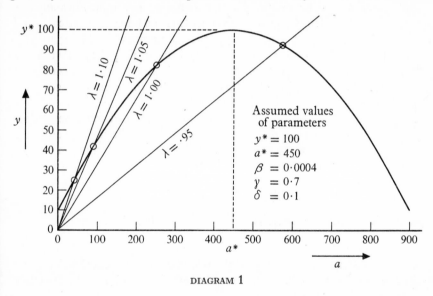

DIAGRAM 1

which c is consequently given by inserting these values of a and y in the third equation of (11).

It can be seen from the diagram that if the assets–output ratio corresponding to maximum output were smaller, that is if the parabola, though constrained by the left side and top of the diagram, was pushed to the left, then higher values of a, y and c could be maintained for given λ. In the limit, in which the parabola coincided with a certain section of the y axis, no assets would be used and an indefinitely large rate of growth would be maintainable.

It can also be seen that the values of a, y and c diminish rapidly for increasing values of λ. Thus the fact that with steady growth a population whose saving behaviour is characterized by (9) will always finance a maintainable level of new asset formation does not mean that it will acquire an adequate, still less an optimal, amount of new assets. These assertions must, however, await the development of a normative theory in section 3.

The first three equations of (11) permit the maintainable value of c to be expressed in terms of λ. The actual values of c and λ which will be maintained can then be found by introducing the fourth equation of (11). In practice there is likely to be only one set of values that need be considered.

3. NORMATIVE THEORY

This section is concerned with optimal states and the contrast between these and the states maintainable under the assumptions made in the preceding section.

An optimal state is here defined in terms of assets per head. Formally it is assumed that the saving behaviour given by (9) is no longer a restriction and that, λ being treated as a parameter, the community is willing to save whatever is needed to meet certain aims of economic policy. Two alternatives, the maximization of output per head and the maximization of consumption per head, suggest themselves and the consequences of a choice between these will now be examined.

In either case the appropriate level of assets per head must be settled. If y is to be maximized the required value of a can be obtained by equating to zero the derivative of y with respect to a from the first equation of (11). Thus

$$\frac{dy}{da} = -2\beta(a-a^*) = 0 \tag{13}$$

whence, as is otherwise obvious,

$$a_y = a^* \tag{14}$$

where a_y denotes that value of a which will maximize production.

In the alternative case, c can be obtained by combining the first two equations of (11). Thus

$$c = y^* - a(\lambda-1) - \beta(a-a^*)^2 \tag{15}$$

and the value of a, a_c say, which will maximize c, given λ, is obtained by equating to zero the partial derivative of c with respect to a in (15). Thus

$$\frac{\partial c}{\partial a} = -(\lambda-1) - 2\beta(a-a^*) \tag{16}$$

whence

$$a_c = a^* - \frac{(\lambda-1)}{2\beta} \tag{17}$$

The parameter β is necessarily positive and so for an increasing population, $\lambda > 1$, a_c is necessarily less than a_y. Thus, since consumption per head is clearly the variable to be maximized, a steadily growing community should not endeavour to reach the maximum point on its production function since by aiming at a lower level of assets and output per head it will be able to enjoy a higher level of consumption per head. Indeed a comparison of (13) and (17) shows that in order to maximize consumption for a given value of λ, assets should be acquired up to the point at which their marginal product is $(\lambda-1)$ and not beyond this to the point at which their marginal product falls to zero.

This part of the theory can be presented in a way which permits comparison with (11). Thus the third equation of (11) disappears and its place is taken by (17). The four variables a, y, c and λ are now related as follows

$$\left.\begin{aligned} y &= y^* - \beta(a-a^*)^2 \\ y &= c+a(\lambda-1) \\ a &= a^* - \frac{(\lambda-1)}{2\beta} \\ \lambda &= \theta + \mu c \end{aligned}\right\} \tag{18}$$

The position now reached can again be illustrated by a diagram. By substituting for a from (14) and from (17) into the first two equations of (11) it follows that, for given λ,

$$c_y = y^* - a^*(\lambda-1) \tag{19}$$

and

$$c_c = y^* - a^*(\lambda-1) + \frac{(\lambda-1)^2}{4\beta} \tag{20}$$

Finally, the maintainable value of consumption, c, is given in terms of a and λ by substituting for y from the second into the third equation of (11) and a is given in terms of λ by equating the right-hand sides of (12) and the first equation of (11). The last relationship is most conveniently found graphically by collating for various values of λ the points at which the straight line crosses the parabola in diagram 1. Hence c can also be expressed in terms of λ. For different values of λ, the values of c_y, c_c and c are shown in diagram 2 below for the values of the parameters used in drawing the first diagram.

As can be seen from a comparison of (19) and (20), for a given value of λ, $c_c \geqslant c_y$, the point of equality being reached when $\lambda=1$. The minimum value of c_c is reached at $\lambda=1+2\beta a^*=1\cdot4$. The relationship of c to λ is complicated and is shown in the diagram over the range $\lambda=0\cdot9$ to $\lambda=1\cdot2$. The value of c rises to that of c_c in the neighbourhood of $\lambda=0\cdot96$ but thereafter falls sharply to roughly between 80 and 50 per cent of the value of c_c over the range $\lambda=1\cdot00$ to $\lambda=1\cdot05$. This means that, insofar as the present numerical example can be considered indicative of the kind of situation met with in practice, saving behaviour of the type assumed might have a seriously adverse effect on the standard of living at those rates of population growth that are actually encountered.

It can also be inferred from diagram 2 that only in the most unusual circumstances, in which certain special relationships subsist between the parameters of the system, will a maintainable state, still less the state actually maintained, coincide with the optimal state.

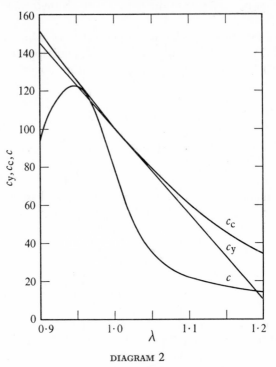

DIAGRAM 2

It may be observed in passing that for a given value of λ, c can be expressed in a simple way in terms of a, a_c, c_c and β. In fact, a_c and c_c are given respectively by (17) and (20). By combining these expressions it can be seen that

$$c_c - c = \frac{[(\lambda - 1) + 2\beta(a - a^*)]^2}{4\beta} = \beta(a - a_c)^2 \tag{21}$$

since

$$a^* = a_c + \frac{(\lambda - 1)}{2\beta} \tag{22}$$

Hence

$$c = c_c - \beta(a - a_c)^2 \tag{23}$$

which may be compared with the production function.

4. POLICY PRESCRIPTION

It must often be the case in practice that actual technological, economic and demographic relationships lead to maintainable states that are not optimal.

To say this is not to imply that human communities are frequently incapable of rational adaptation to circumstances but only that such adaptation may take a considerable time especially for a community whose situation is suddenly changed radically after a long period of stability. Thus if a second production function, with appropriately lower values of a^* and y^*, is imagined to be inserted in diagram 1 it can be seen that the saving function implied in the diagram might lead to a maintainable state that was approximately optimal. This degree of adaptation might, however, only have been achieved after decades or even centuries of gradual adjustment and a sudden change to the higher production function might leave the community temporarily incapable of the new adjustment required.

If λ is treated as a parameter then a better adaptation of the community to a given technological environment involves, in terms of the system of section 2, a different means of determining saving from that which is embodied in (9) with the given values of the parameters. In fact the saving principle enunciated many years ago by Ramsey [131] is designed not only to enable a transition to be made to an optimal state, bliss, but also to ensure that this shall take place at a rate which itself possesses certain optimal properties. As stated by Ramsey this principle applies to the case of a stationary population.

(*a*) *Ramsey's saving principle.* It is assumed that individuals do not discount later enjoyments in comparison with earlier ones, that enjoyments and sacrifices at different times can be calculated independently and added, that no misfortunes will occur to sweep away accumulations and that a given generation need not be deterred by the thought that a subsequent generation might selfishly consume the savings of the past. On this basis Ramsey shows that the flow of saving multiplied by the marginal utility of consumption should always equal the excess of the utility to be derived from the maximum rate of consumption over the utility actually derived from the current rate of consumption. If $u = \phi(c)$ denotes the utility derived from consumption, then this relationship can be expressed in the form

$$s \cdot \frac{du}{dc} = u_c - u \tag{24}$$

where u_c denotes the utility derived from the maximum attainable consumption, that is, from c_c. In order to derive this relationship Ramsey makes use of: (i) a production function in which output depends on the amounts of labour and assets; (ii) the consideration that at all times the marginal disutility of labour should equal the marginal product of labour multiplied by the marginal utility of consumption at the time; and (iii) the consideration that at all times the loss sustained from foregoing an increment of consumption over an infinitesimal period of time should equal the gain derived from the postponement.

5

In order to see the kind of behaviour to which this rule will give rise it is necessary to introduce the relationship between consumption and the utility associated with it. It will be assumed here that as consumption rises its utility will approach a maximum value of η and that there exists a minimum level of consumption, σ, to which the community will not allow itself to be reduced. Thus the utility function might take the form

$$u = \eta - \frac{k}{c - \sigma} \tag{25}$$

Then

$$\frac{du}{dc} = \frac{k}{(c - \sigma)^2} \tag{26}$$

and from (24) and (26)

$$s = \frac{(c_c - c)}{(c_c - \sigma)}(c - \sigma) \tag{27}$$

Thus, on these assumptions, saving should always be a certain proportion of the excess of actual consumption over the minimum level. This proportion is given by the ratio of the amount by which maximum consumption exceeds actual consumption to the amount by which maximum consumption exceeds the minimum level. Also if $c = \sigma$ no saving will take place and the community will not only never get to bliss but will never improve its present minimum level of consumption. If on the other hand $c > \sigma$ the community will eventually get to bliss though the time taken may be very long.

From (1), (2) and (27) it follows that

$$y = \frac{c(c_c - c) + c_c(c - \sigma)}{(c_c - \sigma)} \tag{28}$$

whence, under the conditions of the problem,

$$c = c_c - \sqrt{(c_c - y)(c_c - \sigma)} \tag{29}$$

Thus given an initial value of y, c can be calculated from (29), then s from (27) (or from (1) and (2)) and finally Ea by the addition of s to the opening stock of assets, a. This permits Ey, Ec and Es to be calculated and so on indefinitely.

The situation just described can be illustrated diagrammatically. Suppose that a stationary population the production possibilities of which are characterized by (4) has failed in the past to acquire an assets–manpower ratio of a^*. To take an extreme example, suppose that in the past $a = s = 0$, that $y = c = 10$ and, further, that $\sigma = 0$. Suppose that from period 1 onwards it is decided to remedy this situation according to the prescription just given. Then, in period 1, $y = 10$ as before but c now falls in accordance with (29) to $100 - \sqrt{9000} = 5 \cdot 1$. At the same time s rises to $4 \cdot 9$ which is also the value of Ea. The subsequent paths of the four variables are shown in diagram 3 below.

The variables a, y and c will accelerate for a time and then slow down as they approach their maximum values. The variable s will rise to a maximum of about 25 in approximately 16 time periods and will then decline and approach zero asymptotically.

In this example the initial rate of saving is extremely high, almost 50 per cent of income. This unplausible result is due to the assumption that $\sigma = 0$. If instead it were assumed that $\sigma = 7$, then, in period 1, c, in accordance with (29), would fall only to 8·5 and saving would rise only to 1·5.

It will be observed that, with different initial conditions, a community would still approach bliss along the curves shown in the third diagram.

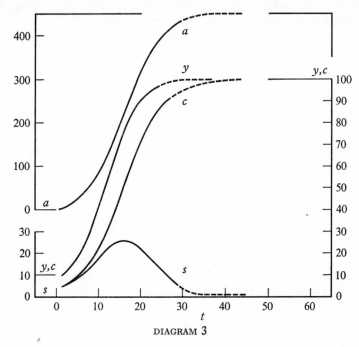

DIAGRAM 3

Thus in the example illustrated in the first diagram a stationary population reaches a maintainable state with $y = c = 81$, $s = 0$, $a = 245$. If, with $\sigma = 0$, this community decides to consume in accordance with (29) it will, in the first time period, set $c = 100 - \sqrt{1900} = 56·4$ and $s = 24·6$. This community is already past the asset–manpower ratio which requires the maximum rate of saving and in subsequent time periods this rate will gradually decline while a, y and c rise.

(b) *A saving principle for a growing economy.* If $\lambda \neq 1$ there appear to be substantial difficulties in deducing an optimal saving principle by a procedure similar to Ramsey's. In these circumstances a useful saving principle can be reached by the following reasoning.

Saving can be divided into two parts the first of which, s_1, is needed to maintain the level of assets per head which has been reached at the beginning of a period while the second, s_2, is needed to increase the level of assets per head towards the optimum. If $\lambda = 1$, then $s_1 = 0$, any saving that is done is available for improving the level of assets per head and its amount can be determined by Ramsey's principle. If $\lambda \neq 1$ then $s_1 = a\,(\lambda - 1)$ and this amount of saving is a prior charge on income if assets are to be maintained. In this case it is only the balance of income, $[y - a(\lambda - 1)]$ which is available for distribution between consumption and an improvement in the level of assets per head. On this basis, and allowing for the difference between the population at the middle and at the end of a period, the system of equations can be developed as follows.

Initially, certain values of y and a have been achieved. Output, y, can be written as the sum, $c + s_1 + s_2$, where

$$s_1 = a(\lambda - 1) \tag{30}$$

and

$$s_2 = \frac{(\lambda + 1)}{2} \frac{(c_c - c)}{(c_c - \sigma)}(c - \sigma) \tag{31}$$

whence c can be expressed as a function of y. Given c, the value of s_2 can be determined and from this Ea, assets at the end of the period, is given by

$$Ea = (a + s_1 + s_2)/\lambda = a + \frac{s_2}{\lambda} \tag{32}$$

If $\lambda = 1$ this system reduces to the one given in the preceding subsection. The state of bliss is reached when $c = c_c$, $s_2 = 0$, $s = s_1 = a_c(\lambda - 1)$ and $y_c = c_c + a_c(\lambda - 1)$. At the other extreme an inability to improve occurs if $[y - a(\lambda - 1)]$ is no larger than σ. For in the case of equality $c = \sigma$, $s_2 = 0$ and $y = \sigma + a(\lambda - 1)$.

(c) *Foreign aid and domestic saving.* If $y - a(\lambda - 1) > \sigma$, it is possible for a community to contribute some saving, however small in amount, for the purpose of increasing its stock of assets, and, by a contribution of this kind, the community will eventually arrive at bliss. Thus if there is any margin for improvement-saving, s_2, a community can in principle always raise its standard of living to the optimal level without the advantage of foreign aid. Such aid is useful for two purposes. First, when a community is living at the subsistence level, to get the whole process of improvement started and, second, to accelerate the rate of improvement. A grant of aid will however alter the situation of a community and, to analyse its effects, it is necessary to distinguish between domestic output, $\dot{y}$, and income, $z = y + g$, where g denotes the amount of foreign aid. The amount to be divided between c and s_2 is now $y + g - a(\lambda - 1)$ and if this is done according to the preceding principles there will be an increase in c as well as in s_2 and a part of the aid will in fact be consumed.

Unless the community is a small one, the amount of aid available is likely to be small compared with the community's own production and also limited in duration. Unless, therefore, special steps are taken to ensure that it shall be used for improvement, in addition to the domestic savings which emerge without it, its net contribution is quite likely to be disappointing. The problem cannot be solved, though it may be eased, by providing the aid in the form of assets and it involves a more direct influence over spending and saving behaviour. Such an influence, however, is required in any case if a community is to exploit its technological environment more fully but does not do this of its own accord.

5. CONCLUSIONS

The situation analysed in this paper and the conclusions to be drawn from the analysis can be summarized as follows:

(*a*) In section 2 it was shown, on the basis of a simple model, that a community whose spending and saving behaviour is characterized by a relationship of the form usually assumed may well reach a steady state in which its standard of living falls short of what could be achieved in the existing technological environment. Generally speaking it seems likely that the progress towards bliss (which is a relative term) will be smaller, the faster the community is growing in numbers and the larger is the amount of assets per head needed to produce the maximum output per head.

(*b*) In section 3 the amount of assets per head needed to maximize consumption per head in a growing community was derived and it was shown to be less than the amount needed to maximize production per head.

(*c*) A comparison of the results of sections 2 and 3 indicates the gap between what has been achieved (and can be maintained) and what might be achieved if the technological environment was more fully exploited.

(*d*) The existence of this gap raises the problem of moving from one state to the other. This problem is considered in terms of a saving principle, originally enunciated by F. P. Ramsey, which will provide the additional saving needed to move the community to a higher point on its production function than the one which it reached and maintained under the original assumptions. This saving principle relates to the case of a stationary population and a similar principle, appropriate to a growing population, is suggested. The role of foreign aid as a supplement to domestic saving is briefly considered.

V

THREE MODELS OF ECONOMIC
GROWTH

1. INTRODUCTION

This paper is concerned with very simple models of economic growth. In micro-economics it is usual to deduce relationships by reference to some maximizing principle intended to reflect rational behaviour. A good example of this is the theory of consumers' behaviour. The formulation of such a theory is quite an achievement but one must recognize that its practical results are often disappointing. This is because it is static, it assumes important variables, like tastes, to be constant, and it is based on a rudimentary and amateurish psychology. In fact the main guidance this theory gives in applied work might almost equally well have been obtained from an empirical approach based on the observation that the consumer responds to changes in income and prices and on the assumption that other influences can be ignored. Market experience, in which individual behaviour is aggregated over the population of consumers, shows that this simple approach contains a good deal of truth and that when it fails, then the theory is of limited value in suggesting useful new avenues to explore. In fact its main value seems to be in suggesting manageable simplifications which will allow price interactions to be studied. But in practice we may have to concern ourselves with such things as the interaction of supply and demand, the phenomenon of saturation, and the time-form of responses to changes in income and prices. In setting up models which will enable us to study these aspects of the actual world, we leave the tidy consistency of static theory and become involved in piecemeal extensions of it which typically do not fit together into a new, more general, harmonious whole. An account of some of the problems encountered in this field is given in [162].

Thus the partial macro-economics of the market or the industry is based on micro-economic theory, which in its turn is based on a maximizing principle. In complete macro-economics, of which trade-cycle theory and growth theory are perhaps the most obvious examples, the position is different. In the first place, just because the models in this field aim at completeness it is more obviously important to put them in a dynamic form. In the second place, since the phenomena studied are so vast, it is usual to start off with aggregated variables such as total consumption or total investment, with the result that attention is fixed on variables which at the outset are distanced through aggregation from individual behaviour. As a consequence theories in this field tend to be relatively superficial and to ignore any optimizing principle which might be supposed to lie at the roots of human endeavour.

With this idea in mind I shall examine three models concerned with economic growth. The first is the familiar multiplier–accelerator model, or its variant, which involves a capital coefficient in place of the accelerator. This is a descriptive model which does not dig very deep. The second is Mahalanobis' two-sector model of economic growth. This traces out the consequence of a particular planning decision, namely, the proportion of investment to be devoted to expanding capacity in the investment goods industry. The consequences of varying this proportion are worked out, but no optimizing solution is considered. The third is Ramsey's model for determining at any time the optimum rate of saving. This can be regarded as a descriptive model based on an optimizing principle, and points the way to a rational solution of Mahalanobis' problem.

All three models are set out in a very simple form. It will be shown that the steady state to which the third model tends is identical with the outcome of the first model but that in addition it traces from the initial conditions to the steady state a transient path which connects some of the basic phenomena of economic development. It will also be shown that the second model tends to the first under certain simplifying assumptions.

2. THE MULTIPLIER–ACCELERATOR MODEL

This model is so familiar and the version of it given here is so simple that it can be written down with very little comment. The first relationship states that saving S is a constant proportion σ of income or product Y. That is,

$$S = \sigma Y \tag{1}$$

The second relationship states that investment V is a constant proportion α of the rate of change of income $\dot{Y} \equiv dY/dt$. That is,

$$V = \alpha \dot{Y} \tag{2}$$

The third relationship is the accounting identity

$$S \equiv V \tag{3}$$

The third equation enables us to equate the right-hand sides of (1) and (2), giving

$$\dot{Y}/Y = \sigma/\alpha \tag{4}$$

a simple differential equation which, on integration, yields

$$Y = Y_0 e^{\sigma t/\alpha} \tag{5}$$

where Y_0 is a constant of integration which is equal to the initial value of Y when $t=0$.

So with this model we have exponential growth, the instantaneous rate of growth, λ say, being equal to σ/α. If $\sigma=0.15$ and $\alpha=5$, we should have $\lambda=0.03$.

If the accelerator α is replaced by a capital coefficient k, then (2) is replaced by

$$\dot{Y} = kV \tag{6}$$

whence, combining (1), (3), and (6), it follows that

$$Y = Y_0 e^{\sigma k t} \tag{7}$$

in which case $\lambda=\sigma k$.

3. A TWO-SECTOR MODEL FOR INVESTMENT PLANNING

In [105] Mahalanobis considers the following problem. In planning, one of the decisions that has to be taken is the allocation of investment resources between the investment-goods industries and the consumption-goods industries. If a large share of investment is allocated to the consumption-goods industries, they will be able to operate initially at a relatively high level. But if this is done, the investment-goods industries will receive only a small share, and so will grow slowly. With an allocation fixed over time, the result will be that the consumption-goods industries will receive a large share of a slowly growing total. In the short run this will lead to a relatively high output of consumption goods, but in the long run the output of consumption goods would be higher if they had had a smaller share in the output of an investment-goods industry which as a consequence of that smaller share would have been able to grow more rapidly. This model is designed to show how these alternative possibilities work out.

Suppose that total income (or product) Y is divided into two parts, the output of the investment-goods industry V, and the output of the consumption-goods industry C. Let the income–capital (product–asset) ratios in these two industries be k_v and k_c, respectively, and let the proportion of investment-goods output allocated in each period to the investment-goods industry be π.

At any moment the rate of investment in the investment-goods industry is πV, and so the rate of change of output of this industry is $k_v \pi V$. Thus

$$\dot{V} = k_v \pi V \tag{8}$$

whence, on integration,

$$V = V_0 e^{k_v \pi t} \tag{9}$$

where V_0 is the initial output in the investment-goods industry. By similar reasoning,

$$\dot{C} = k_c(1-\pi) V$$

$$= k_c(1-\pi) V_0 e^{k_v \pi t} \tag{10}$$

whence, on integration,

$$C = \left\{Y_0 - \left[1 + \frac{k_c(1-\pi)}{k_v\pi}\right]V_0\right\} + \frac{k_c(1-\pi)}{k_v\pi}V_0 e^{k_v\pi t} \tag{11}$$

where the term in {} is the constant of integration consistent with the initial condition that the initial output of consumption goods C_0 is equal to the total initial output Y_0 minus the initial output of investment goods V_0.

Finally, if we add together (9) and (11), we have

$$Y = \left\{Y_0 - \left[1 + \frac{k_c(1-\pi)}{k_v\pi}\right]V_0\right\} + \left[1 + \frac{k_c(1-\pi)}{k_v\pi}\right]V_0 e^{k_v\pi t} \tag{12}$$

We thus see that the output of investment goods grows exponentially, while the output of consumption goods has a constant and an exponential component. If $\pi = 1$, the output of investment goods grows as fast as possible, but from (11) it follows that the output of consumption goods never departs from its initial value C_0. Of course such a high rate of growth in the output of investment goods is useful only if it is intended eventually to lower the value of π when the output of the investment-goods industry has been built up as quickly as possible to what is considered a suitable level.

In this model k_v and k_c are constants and not necessarily equal. This being so, it is natural to inquire whether the planners, or the community, prefer the investment goods or the consumption goods which can be obtained from a marginal unit of investment, and to solve the allocation problem accordingly. Since these preferences are likely to change over time, it might be expected that π would be changed over time. But we will leave this question until we come to the next model.

If we assume that $k_v = k_c = k$, say, and also that the fixed allocation process has always been in operation, then the form of the model is greatly simplified. Thus in place of (9) and (12) we have

$$V = V_0 e^{k\pi t} \tag{13}$$

and

$$Y = Y_0 + \frac{1}{\pi}V_0(e^{k\pi t} - 1) \tag{14}$$

so that in time $V \to \pi Y$ or, in other words, the proportion of income saved and invested tends to the proportion π of investment output allocated to the investment-goods industry.

4. AN OPTIMIZING ALLOCATION MODEL

It is to the undying credit of Ramsey that over thirty years ago he propounded a solution to the kind of problem that has just been posed [131]. Both his model and the technique which it employs, the calculus of variations, have attracted remarkably little attention among economists despite an excellent introduction to both by Allen [4]. The technique in a form suitably modified for computing is only now coming into its own in the work of the dynamic programmers, well exemplified by Bellman [16].

In its simplest form Ramsey's model can be set out as follows. The productive system produces a single good which can be either consumed or invested. The amount produced Y depends on the amount of labour L and capital K employed. Let the production function take the form

$$Y = f(L, K) \tag{15}$$

Consumption C is the excess of production Y over investment $V \equiv \dot{K}$, where $\dot{K}$ is the rate of change of capital. Thus

$$C \equiv Y - \dot{K} \tag{16}$$

The object of economic life is to maximize total utility, which is composed of the utility of consumption minus the disutility of labour. This utility is not just a thing of the moment but stretches out over an indefinitely prolonged future. We should like to know therefore how to allocate product between consumption and investment at every point of time so that the total utility of the system over time should be a maximum.

Let us denote the utility of consumption by M, the disutility of labour by N and total utility by U. Then, assuming that M depends only on C, we can write

$$M = \phi(C) \tag{17}$$

where ϕ is only restricted by the consideration that there is a finite upper limit to M, since economic considerations alone cannot yield an infinite utility. Further, assuming that N depends only on L, we can write

$$N = \psi(L) \tag{18}$$

Accordingly

$$U = \phi(C) - \psi(L) \tag{19}$$

and we wish to maximize an amount W, where

$$W = \int_{t_0}^{t_1} U \, dt \tag{20}$$

In this formulation no account is taken of time discounting in calculating future utilities. This can be done without difficulty, but I shall not do it here.

If we consider the variables on which U ultimately depends, we see that they are L, K, and $\dot{K}$. So we can rewrite (20) in the form

$$W = \int_{t_0}^{t_1} F(L, K, \dot{K}) \, dt \tag{21}$$

Labour and capital can be regarded as functions of time, and our problem is to choose these functions in such a way as to maximize W. The condition for this is that the time paths of labour and capital should be such as to satisfy Euler's equation; that is to say, if X stands for either L or K, then

$$\frac{\partial F}{\partial X} = \frac{d}{dt}\left(\frac{\partial F}{\partial \dot{X}}\right) \tag{22}$$

If we apply this condition to (21), we see that

$$\frac{\partial F}{\partial L} = \phi'(C)\frac{\partial Y}{\partial L} - \psi'(L) = 0 \tag{23}$$

where ϕ' and ψ' are the derivatives of ϕ and ψ with respect to C and L respectively. From (23) it follows that

$$\frac{\partial Y}{\partial L} = \frac{\psi'(L)}{\phi'(C)} \tag{24}$$

or, in words, that at all times the marginal product of labour must equal the ratio of the marginal disutility of labour to the marginal utility of consumption.

In a similar way

$$\frac{\partial F}{\partial K} = \phi'(C)\frac{\partial Y}{\partial K} = \frac{d}{dt}[-\phi'(C)] \tag{25}$$

whence

$$\frac{\partial Y}{\partial K} = -\frac{1}{\phi'(C)} \cdot \frac{d}{dt}[\phi'(C)] \tag{26}$$

or, in words, at all times the marginal product of capital must be equal to the proportionate rate of decrease of the marginal utility of consumption over time.

The time paths of L, K, and C can now be found from the differential equations (16), (24) and (26), the boundary conditions, namely the amounts of labour and capital employed initially and aimed at in the final state, being used in determining the constants of integration. On this basis Ramsey shows that

$$\dot{K} = \frac{U^* - [\phi(C) - \psi(L)]}{\phi'(C)} \tag{27}$$

where U^* is a measure of bliss, the maximum utility of which the economy is capable. Thus saving should equal the excess of bliss over actual utility divided by the marginal utility of consumption. A remarkable feature of this result is that it is independent of the form of the production function except in so far as bliss is determined by limited production possibilities.

Having established this model at rather greater length than the preceding ones, let us now see what happens if we assume simple forms for the various functions, as we did in the earlier models. Let us suppose that the production function (15) is a linear homogeneous function of the first degree; that is,

$$Y = \omega L + \rho K \tag{28}$$

With (28) the marginal product of capital is always equal to ρ. On substituting this value of $\partial Y/\partial K$ into the left-hand side of (26), we obtain

$$\frac{1}{\phi'(C)} \cdot \frac{d}{dt}[\phi'(C)] = -\rho \qquad (29)$$

whence, on integration,

$$\phi'(C) = Ae^{-\rho t} \qquad (30)$$

where A is a constant of integration. Thus, over time, the marginal utility of consumption follows an exponential decay curve.

Now suppose that the utility function (17) takes the form

$$M = \phi(C) = M^* - \beta(C - \bar{C})^{-\theta} \qquad (31)$$

where M^* is the upper limit of M and $\bar{C}$ is the absolute subsistence level below which it is impossible to survive. If we differentiate this equation with respect to C, we obtain $\phi'(C)$, and if we equate the resulting expression to the right-hand side of (30), we obtain

$$\beta\theta(C - \bar{C})^{-(\theta+1)} = Ae^{-\rho t} \qquad (32)$$

whence

$$C = \bar{C} + (C_0 - \bar{C})e^{\rho t/(\theta+1)} \qquad (33)$$

where C_0 is the initial value of C at $t = 0$. Thus we find that the excess of actual consumption over the subsistence level shows simple exponential growth.

Finally, suppose that the disutility function (18) takes the form

$$N = \psi(L) = \gamma L^\eta \qquad (34)$$

If we differentiate this expression with respect to L, we obtain $\psi'(L)$. Combining this result with (24) and (28), we obtain

$$\gamma\eta L^{\eta-1} = \omega Ae^{-\rho t} \qquad (35)$$

whence

$$L = L_0 e^{-\rho t/(\eta-1)} \qquad (36)$$

where L_0 is the initial value of L at $t = 0$. Thus, on the assumptions made here, the quantity of labour, like the marginal utility of consumption, follows an exponential decay curve. Eventually, no labour will be used, since some disutility attaches to it and, with a production function of the form of (15), it is possible to produce as much output as is wanted with capital alone. Accordingly, with this particular model we cannot have reached bliss until we are fully automated.

Before proceeding, let us observe that by combining (32) and (35), we can relate L to C. In fact

$$L = \left(\frac{\beta\theta\omega}{\gamma\eta}\right)^{1/(\eta-1)} (C_0 - \bar{C})^{-(\theta+1)/(\eta-1)} e^{-\rho t/(\eta-1)} \qquad (37)$$

We will choose the initial period $t=0$ as the last moment at which the economy is without capital. This is the last moment at which the whole product is produced with labour alone. Thus

$$Y_0 = \omega L_0 \tag{38}$$

and this relationship when combined with (37) gives, on dividing through by η,

$$\frac{Y_0}{\eta} = \left(\frac{\beta\theta}{\gamma}\right)^{1/(\eta-1)} \left(\frac{\omega}{\eta}\right)^{\eta/(\eta-1)} (C_0-\bar{C})^{-(\theta+1)/(\eta-1)} \tag{39}$$

The purpose of this relationship will be apparent in a moment.

By combining (27), (31) and (34), we can now calculate the rate of saving on the specific assumptions made about the forms of the functions. Thus we obtain

$$\dot{K} = \frac{U^* - M^* + \beta(C-\bar{C})^{-\theta} + \gamma L^\eta}{\beta\theta(C-\bar{C})^{-(\theta+1)}} = \frac{(C-\bar{C})}{\theta} + \frac{Y_0}{\eta} e^{-\rho t/(\eta-1)} \tag{40}$$

from (37) and (39), since at bliss no labour is used and so, there being in those circumstances no disutility of labour, it follows that $M^* \equiv U^*$.

The time paths of Y, C, and $\dot{K}$ can now be written down explicitly. By combining (16) and (40), we find that, for $t=0$,

$$Y_0 = \left[\frac{\theta+1}{\theta} \cdot \frac{\eta}{\eta-1}\right] C_0 - \frac{\eta}{\theta(\eta-1)} \bar{C} \tag{41}$$

whence

$$Y = C_0 \left\{\frac{\theta+1}{\theta} e^{\rho t/(\theta+1)} + \frac{\theta+1}{\theta(\eta-1)} e^{-\rho t/(\eta-1)}\right\} + \bar{C}\left\{1 - \frac{\theta+1}{\theta} e^{\rho t/(\theta+1)} - \frac{1}{\theta(\eta-1)} e^{-\rho t/(\eta-1)}\right\} \tag{42}$$

$$C = C_0\{e^{\rho t/(\theta+1)}\} + \bar{C}\{1 - e^{\rho t/(\theta+1)}\} \tag{43}$$

and

$$\dot{K} = C_0\left\{\frac{1}{\theta} e^{\rho t/(\theta+1)} + \frac{\theta+1}{\theta(\eta-1)} e^{-\rho t/(\eta-1)}\right\} - \bar{C}\left\{\frac{1}{\theta} e^{\rho t/(\theta+1)} + \frac{1}{\theta(\eta-1)} e^{-\rho t/(\eta-1)}\right\} \tag{44}$$

so that

$$\frac{\dot{K}}{Y} = \frac{C_0\left\{e^{\rho t/(\theta+1)} + \frac{\theta+1}{\eta-1} e^{-\rho t/(\eta-1)}\right\} - \bar{C}\left\{e^{\rho t/(\theta+1)} + \frac{1}{\eta-1} e^{-\rho t/(\eta-1)}\right\}}{C_0\left\{(\theta+1) e^{\rho t/(\theta+1)} + \frac{\theta+1}{\eta-1} e^{-\rho t/(\eta-1)}\right\} + \bar{C}\left\{\theta - (\theta+1) e^{\rho t/(\theta+1)} - \frac{1}{\eta-1} e^{-\rho t/(\eta-1)}\right\}} \tag{45}$$

On integrating (44) with $K_0=0$, we see that

$$K = \frac{\theta+1}{\rho\theta}\left\{C_0[e^{\rho t/(\theta+1)} - e^{-\rho t/(\eta-1)}] + \bar{C}\left[\frac{\theta}{\theta+1} - e^{\rho t/(\theta+1)} + \frac{1}{\theta+1} e^{-\rho t/(\eta-1)}\right]\right\} \tag{46}$$

so that

$$\frac{Y}{K} = \rho\left\{\frac{C_0\left[e^{\rho t/(\theta+1)}+\frac{1}{\eta-1}e^{-\rho t/(\eta-1)}\right]+\bar{C}\left[\frac{\theta}{\theta+1}-e^{\rho t/(\theta+1)}-\frac{1}{(\theta+1)(\eta-1)}e^{-\rho t/(\eta-1)}\right]}{C_0[e^{\rho t/(\theta+1)}-e^{-\rho t/(\eta-1)}]+\bar{C}\left[\frac{\theta}{\theta+1}-e^{\rho t/(\theta+1)}+\frac{1}{\theta+1}e^{-\rho t/(\eta-1)}\right]}\right\} \quad (47)$$

The behaviour implied by this model can now be summarized. Before there is any capital, an amount L_0 of labour is used, but as capital is built up, this amount is gradually reduced, as shown in (36), and tends to zero as bliss is approached. By means of saving, as shown in (44), capital grows from zero and tends to infinity. The growth rates of capital, output, consumption and investment all tend to $\rho/(\theta+1)$.

From (45) we can work out the initial and ultimate values of the saving ratio. The initial value, σ_0 say, is obtained by putting $t=0$. Thus

$$\sigma_0 = \frac{C_0(\eta+\theta)-\bar{C}\eta}{C_0\eta(\theta+1)-\bar{C}\eta} \quad (48)$$

which further reduces to

$$\sigma_0 = 1/\eta \quad (49)$$

if $C_0=\bar{C}$, that is, if initial consumption is equal to the subsistence level. Now the parameter η is simply the elasticity of the disutility of labour with respect to the amount of labour performed. Accordingly, the initial saving ratio is low if this elasticity is high and *vice versa*. At the other end of the time scale, the value of the ultimate saving ratio, σ_∞ say, is obtained by letting t in (45) tend to infinity. Thus

$$\sigma_\infty = 1/(\theta+1) \quad (50)$$

Now the parameter θ is the negative of the elasticity of the short fall in actual utility (M^*-M) with respect to the excess of actual consumption over the subsistence level $(C-\bar{C})$. So in the end the saving ratio depends on the attitude to consumption and not on the attitude to work.

In a similar way we can work out the initial and ultimate values of the income–capital ratio from (46). This ratio is initially infinite, since at $t=0$ there is no capital, and tends ultimately to ρ.

Thus in the end this economy tends to a constant growth rate equal to the product of the saving ratio and the income–capital ratio as in the earlier models.

5. SOME COMMENTS ON THE OPTIMIZING MODEL

We have just seen that in the optimizing model the saving ratio and the income–capital ratio tend to constant values as assumed in the first model. But in general these ratios are not constant in the optimizing model but take

on values which change through time according to the values of the parameters and to the form given to the relationships of the model.

The time paths set out in (42) through (47) depend on ρ, θ, η, C_0, and $\bar{C}$.
Let us consider these in turn, beginning with the marginal product of
capital. In the specific version of the model given here this magnitude is
treated as a constant, with the result that in (29) we obtain a very simple
differential equation which we can integrate immediately. Such simplicity
is not essential, but to reach definite conclusions we must know what we
expect to be able to do with labour and capital over the time span we consider.
To the extent that we are wrong about this we shall be mistaken in our
calculation of the best course of action to pursue now.

At this point it is useful to introduce the complication of time-discounting
which would probably make the model more realistic as a description of
actual behaviour though it might be considered irrelevant from the point of
view of investment planning. If people discount future enjoyments, they will
be less inclined to save, since the future fruits of their saving will be less
highly valued than the present equivalent. This consideration can be
introduced into the present model by replacing the marginal product of
capital with its excess over the rate of time-discounting; thus, if this rate is
denoted by ρ^*, then the left-hand side of (26) becomes $(\partial Y/\partial K) - \rho^*$ and the
analysis proceeds as before. In general, $\rho^* < \rho$. If $\rho^* = \rho$, there is no net
advantage in saving and the economy will not add to its existing capital stock.
If $\rho^* > \rho$, the economy will consume its existing capital stock and return to
barbarism.

We come now to the utility function. What can we say about the value of
the parameter θ? Suppose we were to ask the question: if at the toss of a coin
you have an equal chance of a 5 per cent cut or an x per cent rise in consumption, what minimum value of x would induce you to accept this
gamble? After repeating this question with different values of the proposed
cut, we might come to the conclusion that θ was of the order of 2. In this case
the community should ultimately save one-third of its income. If we look at
statistics of national income and saving, we find that few countries save at
anything like this rate and that it is certainly not the rich ones that tend to
do so.

Probably the explanation is to be sought in the phenomenon of time-
discounting which, whatever its relevance may be for planning purposes, is
certainly a factor that tends to reduce individual rates of saving. It might
also be supposed that as people get richer, their conception of what constitutes a subsistence level is revised upwards. This, however, would change
the ultimate value of the saving ratio only on the unlikely assumption that
there is no absolute subsistence level, that is, $\bar{C} = 0$. For in this case we have,
writing $\bar{\bar{C}}$ for the conventional subsistence level,

$$\bar{\bar{C}} = \mu C \tag{51}$$

so that the utility function takes the form

$$M = M^* - \beta[(1-\mu)C]^{-\theta} \tag{52}$$

and the saving ratio tends to

$$\sigma = \frac{1}{1 + \dfrac{\theta}{1-\mu}} \tag{53}$$

If on the other hand $\bar{C} \neq 0$, then we must define $\bar{\bar{C}}$ as

$$\bar{\bar{C}} = \bar{C} + \mu(C - \bar{C}) \tag{54}$$

in which case

$$\begin{aligned} M &= M^* - \beta[(1-\mu)(C-\bar{C})]^{-\theta} \\ &= M^* - \beta^*(C-\bar{C})^{-\theta} \end{aligned} \tag{55}$$

which is identical with (31) except that β is replaced by β^*.

This brings us to the disutility function. The measurement of η poses a similar problem to that of the measurement of θ and might be accomplished by a similar type of psychological experimentation. With the particular model adopted here one would expect η to be rather large, say between 10 and 100. From (48) we see the importance of an absolute subsistence level. We have seen that if $\bar{C} = C_0$, then $\sigma_0 = 1/\eta$. But if $\bar{C} = 0$, it follows from (48) that

$$\sigma_0 = \frac{\eta + \theta}{\eta(\theta + 1)} \tag{56}$$

so that even with $\theta = 5$ and $\eta = 100$ we should have $\sigma_0 = \frac{1}{6}$, which seems impossibily high for the initial saving ratio.

6. CONCLUSIONS

From this short treatment of various growth models the following conclusions may be drawn.

1. We have contrasted the simple multiplier–accelerator model with a simple optimizing model and shown that, on very simple assumptions about the forms of the relationships, the optimizing model tends in its consequences to those which follow at all times from the multiplier–accelerator model.

2. The multiplier–accelerator model represents a form of steady-state dynamics which is of limited interest because it summarizes the complicated effects of human behaviour and of technology in terms of constant parameters. By contrast, the optimizing model gives an account of the transient behaviour between the initial and the ultimate states of the system.

3. The optimizing model provides another example of the importance of defining and measuring utility. The static theory of consumers' behaviour is

organized around the concept of utility by means of the technique of un-determined multipliers. In the optimizing model we have the dynamic analogue which makes use of the technique of the calculus of variations.

4. In order to replace the *ad hoc* relationships which figure so largely in macro-economics, the optimizing model tries to penetrate more deeply into the springs of human action. For example, it provides a form of the saving function to replace the *ad hoc* assumption that saving is a constant proportion of income at all times. Thus the psychological motives of economic behaviour are brought into the picture. My version of Ramsey's model does this in a very crude way, but it should be enough to illustrate the unity of the social sciences. The limitations of pure economics as opposed to empirical economics are now becoming generally recognized, and it is high time economists took seriously the psychological and social postulates of their theory instead of squandering so much energy in preserving its purity, which beyond a certain age is just another word for sterility.

5. The optimizing model points the way to a rational solution of the problem posed by the planning model. This indeed is a central problem of political economy: the proper distribution of consumption between the present and the future. If the present is given too much, the economy will grow slowly, or even fail to grow at all, and people will become restive once they have abandoned a fatalistic outlook. If the future is given too much, the present generation, once its revolutionary fervour has abated, will come to question a regime which seems to treat them as peons. Either way the politician had better be on his guard.

6. Finally, it cannot be too strongly emphasized that the models in this paper are all three rather primitive and that the conclusions might be modified if one changed the forms of the relationships used. Such matters, however, can rarely be settled at the theoretical level. The model-builder must choose among acceptable alternatives those which not only are amenable to statistical analysis but also fit the facts. This, of course, is not to suggest that all we need are more and better observations. To suppose that the facts themselves will lead us to a clear and unequivocal system of relationships in a matter as complicated as an economic system is as absurd as to suppose that by purely theoretical processes we can discover a system which will turn out to square with all the facts. Only by going back and forth between theory and observations can we hope to progress in an empirical science like economics.

VI

MODELS OF THE NATIONAL ECONOMY FOR PLANNING PURPOSES

1. INTRODUCTION

In my talk at the Cheltenham Conference I described the computable model of the British economy on which a group of us is working at the Department of Applied Economics in Cambridge. This model and its results are being described in detail in a series of publications entitled *A Programme for Growth*, of which the first paper [30] gave a general description of the model and the second [31] set out the accounting framework for the base year, 1960. Future issues in the series will appear, we hope, at roughly quarterly intervals.

Rather than go through all this again, I propose in the present note to elaborate certain points that arose in the discussion of my talk. There seem to be two questions that particularly need clarification: first, can we place a model like the one I described in its general intellectual, political and administrative environment so that we can see how it is related to the active forces in the real world; and, second, what specific contributions could operational research teams make to it or it make to the objectives of operational research?

2. A SLICE OF LIFE

Every model is an abstraction designed to analyse a slice of life with the object of understanding it better and, if it is at all controllable, adapting it better to our wishes. If we want to succeed in our aims we must not only try to build a good model but we must also try to understand how it fits into its environment. If we fail in this second endeavour, an otherwise promising idea may fail to make headway because it does not seem worth taking up and improving to the people who would be affected by it; of these the inventor is only one of many.

As far as I can see, the relationship of a model to its environment can best be represented by a closed-loop diagram. I shall now give one appropriate to a model of the national economy, though I believe that an essentially similar situation exists for all models that are set up for practical purposes. The diagram is highly schematic and I shall not try to introduce all the finer shades of a complicated situation.

If we want to study economic growth, we must decide on a relevant set of variables and provide a theory which shows how they are related. This theory must be dynamic but it must not be a complete theory with fixed parameters, since if it were we should only discover the one growth rate of which the system was capable. This is the situation with von Neumann's and similar models of the productive process [59] which give rise to a single efficient growth rate: if we want something better there is nothing we can do about it. Such models are valuable for the insights they give into the productive process but not for policy purposes, since it is unlikely that the real world is as simple and rigid as the models.

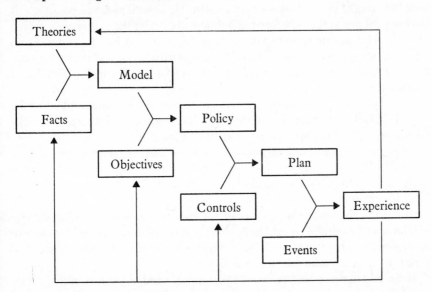

If we combine our theory with a set of observations we can give numerical values to its parameters and so we obtain a quantitative model. The econometrician usually tries to extract estimates of parameters from the statistical analysis of past observations. He usually realizes that many of these parameters are likely to be changing and tries to set up his equations in such a way as to get some information about the rates of change. He may do quite well with this method because economic relationships have a considerable amount of inertia. But changes are certain to take place in the future that cannot be inferred from the past. If he can detect these cases he should try to ask the man-who-knows what form the changes are likely to take. This will immediately take the model builder outside himself and I shall return to this point in the next section.

By combining a model with a set of objectives we obtain a policy, by which I mean a detailed statement of what must be done to realize the objectives but not how these acts are to be carried out. The sort of objectives I have in mind are the rate of growth we should like to achieve or the size of the education or road-building programme that we should like to see.

The objectives with which we start may be impossible to realize or incompatible with one another, and the model can throw light on this so that eventually we stop crying for what a few years ago would have been called the moon. After we have cleared up these points there will still be conflicts of objectives, the most obvious being the conflict between faster growth and old habits or between more goods and more leisure. On such matters the model builder is just like anyone else: his vote counts for one and only one. His job is to point out the difficulties and incompatibilities and to suggest how they might be overcome or reconciled. He should as far as possible put the range of possibilities before the public and then let them choose, entering into the final debate only to point out *non sequiturs*, if he can. There is no reason to suppose that model builders are particularly good policy makers. Even so they perform an essential service; as Marx said, freedom is to know necessity.

I have gone on so far as if objectives were completely independent of the model; in fact quite a number of them are likely to be built into it almost subconsciously. For example, in the Cambridge model it is assumed that consumers, given the price situation, are free to decide the composition of their expenditure, though the model could examine the effects of radical changes in the tax system which would alter relative prices and consequently the way in which consumers would want to lay out their money. Again, at a later stage we intend to elaborate our production relationships so as to optimize the use of labour and capital. I should hope that most people would consider these to be desirable features that ought to be built into any model, though there will certainly be some who take a different view.

If we stop at this point we can see that models and objectives, like theories and facts, are not independent of one another. In theorizing it is useful and often desirable to work with concepts that no one can observe or measure, but in the end the theory must say something about observable facts. A good example in economics is the concept of utility which, until recently, nobody had even tried to measure. Yet it is the central concept of a great deal of economic theory. In the static theory of consumers' behaviour it is assumed that the individual consumer has a preference system for the different goods available and that he tries to maximize the utility of what he buys subject to the fact that goods cost money and he has only a limited amount of money to spend. The solution of this constrained maximum problem gives the individual's demand functions, that is, connects the quantities he buys with his income and the price structure. By adding up individual demand functions and making certain simplifications we arrive at market demand functions which can be tested against observation [168]. Or, again, the classical method of solving the problem of the optimum rate of saving is to maximize the total utility yielded over time by an economic system when the time span is made indefinitely long. This is a problem in the calculus of variations whose solution yields statements about the desirable level of observable magnitudes such as the ratio of saving to income [V]. All such theories would be useless if they said nothing about observable things because we could not test them: our only test of a theory would be its internal consistency. Conversely, facts must be organized in the light of theories. A great many distinctions we make in presenting facts are in no sense natural and in some cases are very difficult to make, but we make them because unorganized facts dissolve into mere sensations which are quite useless for scientific purposes.

The interaction between models and objectives is similar. A model may embody certain objectives, as when some optimizing procedure is built into it; or the objectives may appear as constraints on acceptable solutions; or the model may be used to examine the possibility of reaching certain objectives and the problems to which the attempt to do so would give rise.

It is difficult to specify objectives precisely, partly because it is hard to envisage states of the economy which are outside our experience, partly because there are inevitable conflicts of interest between different groups in the community, and partly because there is no obvious point at which to stop in listing objectives. As we widen the range of those we consider explicitly so we must make sure that our model continues to be relevant. For example, in the Cambridge model, expenditure on education, both current and capital, is a part of final demand and is assumed to be given. We can examine the demands put on the productive system of a large or a small educational programme, but we cannot even determine how much of this programme should relate to the technical education required to provide the different skills which the economy will need in the future. No doubt such a link could be built into the model but it is not there at present.

A similar situation exists with housing which is given simply in terms of a total number of houses to be built and equipped. Since the model contains as yet no regional dimension, it can say nothing about where these houses are to go. Nor can it say anything about restrictions imposed by the wish to preserve rural amenities or by the requirements of urban renewal. For these purposes a very considerable extension of the model would be needed.

If a model is to be useful for planning purposes it is essential that it should contain some degrees of freedom and not specify the relationships connecting the variables so completely that there is no room for manoeuvre. The solution of a determinate model would fix the value of each variable exactly, including, for example, the one rate of growth which was compatible with all the relationships. This would be no good at all; we must be free to see what could be made to give if we wanted to achieve a higher rate. In other words, the purpose of a planning model is to change the patterns of behaviour, not to perpetuate them.

This part of the discussion is intended to show that while a model is indispensable for planning purposes, it deals only with a slice of life. It does not produce *the* answers but a starting point for reaching them.

To continue with the diagram, a combination of objectives and a model yields a policy. A policy tells us what to try to do, not how to do it. Accordingly we must combine a policy with a set of practical measures for carrying it out. These I have called controls. The controls might consist of discussing the plan with the leaders of management and labour in different industries, adapting it to meet their criticisms as far as these seemed reasonable and then leaving them to carry out their part of the plan as they thought best. They might consist of this arrangement supplemented by some additional rules or bits of administrative machinery that would help in this process. Or they might consist of a central authority that took all administrative decisions. In my opinion the liberal countries tend to put too much faith in decentralization without considering what they are trying to do, and the communists tend to think that just because they have a clear idea of what they are trying to do it does not much matter and may even be a good thing if the people who have to do it are tied up from head to foot in red tape.

In simple cases, like the domestic heating system controlled by a thermostat, the administration can be built into the model. But if someone lights a fire under the outside thermometer, the rooms will always be cold until the householder takes the administration into his own hands. In the economic world there are too many people lighting fires under thermometers to enable us to rely on purely automatic devices for regulating the economy, though such devices, properly supervised, might greatly help the administration.

We have now seen the roles of the scientist, the policy maker and the administrator in realizing a plan. Each is necessary and each must learn to work with the others. But each is conscious of his own power position and all too often would like to cut out the others. In limited fields this may work, but, in general, it will not: *si monumentum requires, circumspice.*

The plan is not the end of the road: in a sense it is only the beginning. The plan, when put into practice, comes up against events. The result is our experience of the slice of life with which the plan is concerned. This experience is likely to modify every component part of the plan. Take administration: in most liberal countries the price system is regarded as a good regulator of economic life; but it can be seriously distorted by monopoly power, and it tends to do well only in short-term, clear-cut situations. A free market for cabbages may succeed, says the theory; but the whole history of restriction schemes for primary commodities shows how much the theory leaves out. The attempt to obtain urban amenities through the market mechanism is virtually impossible.

Or, again, take objectives. In the beginning we are likely to over-emphasize immediate and tangible objectives compared with remote and intangible ones. The results of trying to get what we thought we wanted may well change our ideas as to what we really want.

Finally, theories and even facts are also likely to be affected. I do not mean, of course, that saving will cease to equal investment or that we shall come to believe that the average weekly earnings of cowmen in England and Wales in the summer of 1961 were significantly different from £13. But we are sure to find that some of the concepts and relationships which we habitually use are unsatisfactory and can be improved; that some of the facts, which are really expectations of future facts, can be better estimated; and that the scope of the model can usefully be increased.

Thus in the end we revise our definition of the slice of life to which national economic planning relates, and so revise our models, policies and plans. But unless we want anarchy we must not deny the existence of slices of life; nor must we think that the whole cake is our slice. At either of these extremes we can organize nothing: in the first case because we do not admit that there is anything to organize; in the second, because the organizational problem is beyond our powers.

3. OPERATIONAL RESEARCH AND MODELS OF THE NATIONAL ECONOMY

In the preceding section I tried to set a model in its environment, and from what I have said it will be obvious that I do not regard model building as an ivory-tower occupation. So far I have only considered the scientist in relation to the policy maker and the administrator; now I want to consider him in relation to other scientists with connected fields of interest.

The operational researcher is such a person. I assume that one of his many activities is to make a model of his firm or trade which will help to run it more efficiently. If this is correct, I see a two-way traffic between operational research and the building of national economic models. These exchanges should go on at both the model-building and the model-using stage. Let us now look at various areas in which co-operation might be fruitful.

(a) *Taxonomy*. Definitions and classifications are an essential feature of model building at any level. For example, how should we classify the establishments or, if we take a more detailed view, the processes whereby we define an industry? The man who tries to build a model of the whole economy is bound to start with the standard industrial classification and will probably adopt a fairly high degree of aggregation: at present the Cambridge model has only thirty-one industries. When, at a later stage, a more detailed classification is being considered, it would be extremely helpful if those working on models of specific industries could suggest the best subdivisions to adopt within their industries. Of course, to be useful such suggestions would have to take account of the availability of data.

(*b*) *Basic data.* The builder of national models starts off by seeing what he can do with existing official statistics. Though great strides have been made in recent years there are limitations to this source, one of which is that many of the data are not up to date. For example, a full census of production is of the greatest importance, but one is taken only every four or five years and several years are allowed to process and publish it. The derivation of secondary statistics from this primary source calls for further work and involves further delays. Thus the latest official input–output data relate to 1954 and were published in 1961 [174]. For model-building purposes it is necessary to bring these figures up to date and project them into the future. The supplementary data needed for this operation are limited. Accordingly, the model builder has to resort to mathematical methods based on assumptions. He can probably do quite well in this way as far as the recent past is concerned [158]. But when it comes to the future, his position is much more difficult because past trends cannot cover new developments. Yet these new developments undoubtedly cast their shadow in front of them and must be known to operational research teams in at least some of the industries concerned. Information about them would be very useful.

(*c*) *Relationships.* As I said above, the builder of national models tries to estimate the parameters in the relationships of his model from past data, allowing as far as he can for the changing character of these relationships. For example, he will not use the input–output coefficients of 1954, nor even of 1960, but will try to project these coefficients so as to represent the technology of the future. On this subject an operational research team that has studied a particular industry should be able to do a much better job for that industry than the builder of national models can. Again, their opinion would be very useful.

The main difficulties in this area are either taxonomic or arise from differences of purpose. Probably no model builder at the firm or trade level ever deals with an industry as broad as food processing, chemicals or iron and steel; he deals with a subdivision, and often a small subdivision, of one of these categories. Nevertheless, the builders of such models may well want to set them in a wider framework; there should be a point of contact somewhere around the industry level, its exact position depending on the circumstances of each individual case.

So much for model building; let us now turn to the possibilities of co-operation in model using. Here there seem to be two main areas.

(*d*) *Reviewing results.* For any given set of initial assumptions, a national model will produce a vast mass of numerical results relating to a future year. For each industry the model will give:

(i) total output (at base year prices);
(ii) the distribution of this output to different intermediate and final users;

(iii) the cost structure of this output in terms of primary and intermediate inputs; and

(iv) the shadow price of each industry's output.

With this information it is possible to set up a social accounting matrix for the future year. These results are internally consistent from an accounting point of view but still need to be scrutinized very carefully. The reason for this is that inaccuracies in the model may have surprisingly large repercussions in certain limited areas and that coefficients in different parts of the model, being estimated independently of each other, may lead to implausible relationships. For example, in the absence of special knowledge, most of the future input–output coefficients are estimated by projecting recent input–output trends. This may imply that product A will continue to be substituted for product B. But when we look at the future shadow prices of the two products we may find that the price of A has not continued to fall in relation to the price of B. *Prima facie*, therefore, we should not expect the substitution to continue, and we must try to find out whether there is some reason for it other than price or whether we are wrong in assuming that it will continue. In reviewing points such as these, the advice of people familiar with the problems of particular industries would be very valuable.

(e) *Linking models.* The builder of national models has the problem of getting his industry detail right and the builder of industry models may have the problem of setting his detailed model in the national framework. If the two models can be given a point of contact, it should be possible to link them. One way to do this would be to substitute a set of trades, processes, activities, or whatever they should be called, for the industry that contains them in the national model. The national model would then produce results for these trades which fitted into the national picture. These results could then be used as data in the industrial model. This might lead to a revision of the parameters in the national model since it would be difficult to ensure initially that the two models were consistent at their point of contact. However this might work out, the national model would almost certainly be improved and the scope of the industry model would be extended.

4. CONCLUSIONS

Nowadays, in contrast with the position a generation ago, economic model building is going on at all levels, from the firm to the national economy. In all kinds of ways the scientists working at these different levels could increase the efficiency of one another's work. The first thing to do is to open up channels of communication; the next is to get some information flowing in these channels in spite of the ever-shut valves of government and business secrecy. We shall not get very far in this, however, if we always refuse to understand why the valves are shut, or if we think that the man whose hand is on a particular valve is necessarily the motivating agent.

The model builder who wants to see his models put to practical use commits himself to a life of working with others. The idea that economic models can be useful is a comparatively new one and is resisted by many people from some vague apprehension that, once accepted, the models would begin to run their lives. Such people do not see the models as tools which would enable them to do their political, administrative or productive job better, with a consequent increase of satisfaction to themselves as well as to others. Sometimes model builders, in their enthusiasm, seem to fan these fears instead of allaying them; an example of this is the controversy in the early 1950's about 'push-button planning', which brought American official work on the construction of national input–output tables to a halt in the spring of 1953. The fearful ones do their best to misrepresent the model builder and his work, and he should contest this misrepresentation very sharply. At the same time he should not allow himself to be goaded by those emotionally opposed to models into denying the importance of every consideration that he has not allowed for in his model. He should remember that 'those who live in glass houses should not throw stones' or, in the present case, those who promote changes by an appeal to reason should not assume that all their adversaries are unreasonable. Many may be on the lunatic fringe, but not all. Unless the model builder sees himself and his model as part of the real world, in an environment made up of other people and other specialisms, he will soon join the lunatic fringe too.

VII

A DEMONSTRATION MODEL FOR ECONOMIC GROWTH

1. INTRODUCTION

This paper is a sequel to [152, 156] and describes some of the latest developments in the computable growth model on which my colleagues and I are working. The earlier version of this model did not allow the outcome to be affected by changes in relative prices, and so the main feature of the version described below is the juxtaposition of a set of demand functions which are sensitive to relative prices and a set of production functions which are sensitive to relative factor rewards. The connecting link is an expression relating the consumption and output vectors for given growth rates in the components of consumption.

The new model is set out on the diagram overleaf, and will be described bit by bit in the following sections. For ease of presentation a number of realistic details are omitted but I shall try to show, in section 8, that these omissions are more apparent than real and that the required degree of realism can be achieved with the relatively simple categories used in this paper.

2. THE CONSUMPTION FUNCTIONS

In this exposition final demand is composed of consumption and net investment. The consumption vector is related to total expenditure on consumption and to the price vector by a simple system of equations which I have elsewhere [148] called a linear expenditure system, because the expenditure on each commodity is a homogeneous linear function of total expenditure and all the prices. Written as an expression for the quantities demanded, this system takes the form

$$e = \hat{p}^{-1}[\mu b + (I - bi')]\hat{b}^* p \qquad (1)$$

In (1) e is a vector of quantities demanded, p is a vector of prices, μ is total expenditure on consumption, b and b^* are vectors of parameters, and I and i are respectively the unit matrix and the unit vector. Vectors are written as column vectors, a prime denotes transposition and a circumflex accent denotes a diagonal matrix constructed from the vector it surmounts. As was shown in [148], (1) is based on the following three properties:

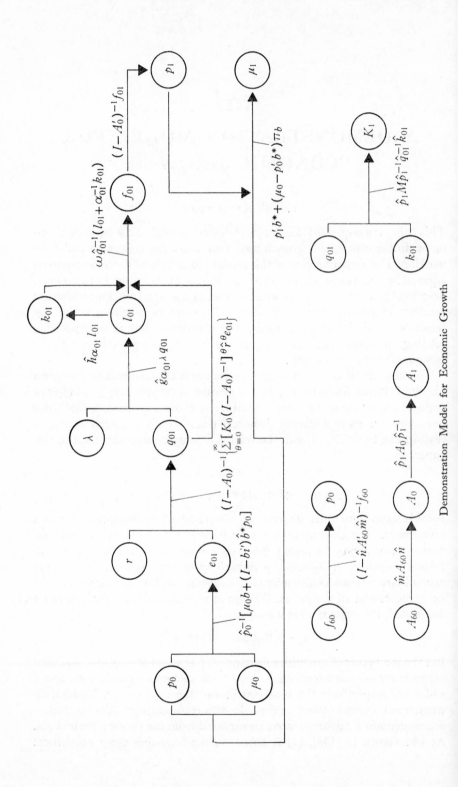

Demonstration Model for Economic Growth

(i) $p'e \equiv \mu$, which is satisfied provided that $i'b=1$; (ii) there is no money illusion since the elements of e are of degree zero in μ and in the elements of p; and (iii) the Slutsky condition is satisfied, that is the matrix of elasticities of substitution is symmetric. Further, as was shown in [85], (1) gives rise to a constant-utility index-number of the cost of living, that is to say one can derive an expression for the amount of money needed in situation 1 to keep the system on the same utility surface as it was on in situation 0, after the price vector has changed from p_0 to p_1. If this amount of money is denoted by μ_1, then

$$\mu_1 = p_1'b^* + (\mu_0 - p_0'b^*)\pi_b \qquad (2)$$

where π_b is a geometric average of the price ratios (period 1 divided by period 0) with the elements of b as weights.

The meaning of the parameters b and b^* can be seen more clearly if (1) is premultiplied by $\hat{p}$ and the terms on the right-hand side are rearranged to give

$$\hat{p}e = \hat{p}b^* + (\mu - p'b^*)b \qquad (3)$$

From (3) we see that each of the elements of the expenditure vector, $\hat{p}e$, is made up of two parts: (i) the purchase of a certain fixed quantity, the relevant element of b^*, and (ii) the allocation of a certain fixed proportion, the relevant element of b, of the excess of the money available, μ, over the cost of all the fixed expenditures, $p'b^*$. Thus the amount of money required for the community to stay on the same utility surface is equal to the new cost of all the fixed expenditures, $p_1'b^*$, plus the supernumerary income of the base period, $(\mu_0 - p_0'b^*)$, multiplied by a geometric index-number of the price ratios, π_b.

We also see from (3) that (1), which is designed primarily to represent the behaviour of private consumers, can be adapted without much difficulty to represent the more or less price-insensitive behaviour of government consumers. Thus if the quantities of education, health facilities, defence, etc., are represented by elements of b^* and if the corresponding elements of b are put equal to zero, consumers as a whole will demand these quantities independently of the price situation and, if prices change, consumers as a whole must be given enough money both to pay for these fixed commitments and to enable private consumers to remain on a given utility surface.

Equation (1) has been adopted for the consumption functions because it combines convenient theoretical properties with modest requirements in the number of parameters to be estimated. It can be modified in numerous ways: trends can be introduced into the parameters to allow for gradual changes in tastes [149], the formulation can be made dynamic [162] and so on. There is some evidence [162] that, at least in its dynamic form, (1) accounts reasonably well for variations in the components of private consumers' expenditure. But, whatever refinements are introduced, (1) has certain limitations: (i) it describes a world of substitutes, and cannot accommodate commodity groups which are complementary or inferior; (ii) the Engel curves connecting elements of e with μ for fixed p are linear though we know [2, 127, 155] that this can only be regarded as an approximation; and (iii) the ordinary price-quantity demand curves connecting each element of e with the corresponding element of p for fixed μ and for the remaining elements of p are hyperbolae. Given that the elements of b lie between 0 and 1 these curves cannot be elastic.

Even in a model in which commodities are grouped in twenty or thirty categories it may turn out that these limitations rule out (1) as an acceptable formulation of consumers' behaviour. If so it will be necessary to give up the theoretical advantages of (1) and work with a set of demand equations that do not possess its properties of uniformity and consistency. In this case the constant-utility index of the cost of living will disappear and we shall have to be content with an approximate calculation of the amount of money needed to achieve a given level of real consumption as prices change. But this should not present serious practical difficulties since the price index-number of (1), μ_1/μ_0, is very like other price index-numbers; all that would really be lost is a certain convenience and tidiness in the algebra.

3. CURRENT OUTPUT REQUIREMENTS FOR GROWTH IN CONSUMPTION

The principal accounting identity of this model expresses two facts: (i) total product is made up of intermediate product and final product; and (ii) final product is made up of net investment and consumption. This identity can be written as

$$q = Aq + v + e$$
$$= (I - A)^{-1}(v + e) \tag{4}$$

In (4) q is a vector of output, v is a vector of demands on the different industries for net investment, e is a vector of demands on the different industries for consumption and A is the current input–output matrix. The vector e here is of course the same as the vector e in (1) which means that in the present exposition a single classification of product is being used, and that the problems of establishing a classification converter [XVI] are ignored.

Net investment, v, is assumed to be related to the excess of next year's output over this year's output, Δq, by a matrix K of capital coefficients. The typical element, k_{st} say, of K represents the amount of products of industry s required as capital goods by industry t to enable it to increase its output by one unit. On this assumption

$$v = K\Delta q \tag{5}$$

Let us now suppose that each element of e is to grow in the future at a certain exponential rate represented by the element of a vector r. Then next year's consumption vector, Ee say (where $E \equiv 1 + \Delta$), is given by

$$Ee = (I + \hat{r})e \qquad (6)$$

If we combine (4), (5) and (6) we find [157] that

$$q = (I - A)^{-1} \left\{ \sum_{\theta=0}^{\infty} [K(I-A)^{-1}]^\theta \, \hat{r}^\theta e \right\} \qquad (7)$$

The term in { } in (7) is the same as $(v + e)$ in (4). The first element of this term (corresponding to $\theta = 0$) is e; the remaining terms (corresponding to $\theta = 1, 2, \ldots$) are, in total, the demands for net investment which arise this year to provide for the growth in the elements of e. Equation (7) cannot be reduced to a matrix-multiplier expression because $[K(I-A)^{-1}]^\theta \hat{r}^\theta \neq [K(I-A)^{-1}\hat{r}]^\theta$ unless the elements of r are all the same. However since the largest element of $K(I-A)^{-1}$ (the direct and indirect demands for capital equipment placed on industry s by industry t to allow industry t to increase its output by one unit) is about 5 and since the average element of r is about 0·05, it follows that the multiplier of e in (7) for $\theta \geqslant 4$ is negligible. Accordingly, q can be calculated from (7) by taking not more than the first four or five terms in the infinite series.

4. THE PRODUCTION FUNCTIONS

The production functions used in the model are of the modified Cobb-Douglas type recently proposed by a number of writers [7, 123]. The principal refinements are two: (i) the elasticity of substitution between labour and capital, though fixed, need not be unity, and (ii) technical progress, represented by an exponential trend, enters along with labour and capital into the function for each industry. As shown in [7] such a function can be written, for industry s, in the form

$$q_s = a_s[(1-b_s)l_s^{-c_s} + b_s k_s^{-c_s}]^{-c_s^{-1}} \qquad (8)$$

In (8) q_s is output, l_s is labour used and k_s is capital used; a_s is a parameter associated with the efficiency with which the primary inputs are used, and is assumed to grow with time, b_s is a parameter associated with the shares of labour and capital in net output, and c_s is a parameter associated with the elasticity of substitution between labour and capital. This elasticity is equal to $(1+c_s)^{-1}$.

If we differentiate (8) with respect to k_s and l_s and take the ratio of the derivatives we obtain the ratio of the marginal physical products of capital and of labour. In a competitive economy these ratios should be the same in each industry. On this assumption

$$k_s = \left[\frac{\alpha b_s}{1 - b_s} \right]^{(1+c_s)^{-1}} l_s \qquad (9)$$

where α is the common ratio of the marginal physical products. If we substitute for k_s from (9) into (8) we obtain

$$q_s = a_s \left\{ (1-b_s) + b_s \left[\frac{\alpha b_s}{1-b_s} \right]^{-c_s(1+c_s)^{-1}} \right\}^{-c_s^{-1}} l_s \tag{10}$$

or

$$l_s = a_s^{-1} \left\{ (1-b_s) + b_s \left[\frac{\alpha b_s}{1-b_s} \right]^{-c_s(1+c_s)^{-1}} \right\}^{c_s^{-1}} q_s \tag{11}$$

Given the output vector, q, whose elements are q_s, and a labour force, $\lambda = \Sigma_s l_s$, it is possible, by an iterative calculation, to determine a value of α. The link between l and q is shown on the diagram in the form

$$l = \hat{g}_{\alpha\lambda} q \tag{12}$$

where the elements of g, which depend on α and λ, are the optimal labour–output ratios in the different industries.

Once the optimal allocation of labour is known, the associated capital stocks follow at once from (9). This link between k and l is shown on the diagram in the form

$$k = \hat{h}_{\alpha} l \tag{13}$$

where the elements of h, which depend on α, are the optimal capital–labour ratios.

5. THE DETERMINATION OF PRICES

The price vector p follows almost at once from the production functions. The wage rate, ω say, being given as a unit of account, the factor costs per unit, f say, are

$$f = \omega \hat{q}^{-1}(l + \alpha^{-1} k) \tag{14}$$

since α is the common ratio of the reward of a unit of labour to the reward of a unit of capital. Since all inputs are primary or intermediate, it follows that

$$p = (I - A')^{-1} f \tag{15}$$

6. THE SOLUTION OF THE MODEL

If the price vector calculated from (15) happened to be the same as the price vector we assumed in determining e in (1) we should have a solution, since in this case it follows from (2) that $\mu_1 = \mu_0$. But in general we may expect that the two price vectors will be different, in which case we can use (2) to calculate μ_1, that is the money to be spent on consumption with the price vector, p_1, given by (15) to maintain the utility level originally intended by μ_0. The replacement of μ_0 and p_0 by μ_1 and p_1, and the attendant changes in the parameters that the change in the price vector will entail, lead to a further cycle of calculations which must be repeated until the price vector settles down to a constant value.

What has just been said is subject to one qualification, namely that the parameters of the model do not change over the cycle of calculations. In order to show more clearly what is involved, both variables and parameters are provided in the diagram with a suffix: 0 if they are values fed in initially; 01 if they are determined in the course of the cycle; and 1 if they emerge at the end of the cycle. Variables and parameters with no suffix, such as λ and r, are assumed fixed and do not change as a consequence of the calculations. If $p_1 = p_0$ then $A_1 = A_0$, but even in this case it does not follow that $K_1 = K_0$ because K_0 is essentially a trial value which is likely to be modified as soon as the required outputs are known. The modification will not, in all probability, be very large but it may be sufficient to justify a further cycle of calculations.

7. THE DETERMINATION OF PARAMETERS AND INITIAL VALUES

It is a feature of the model that it is made as little dependent as possible on past values of the parameters and that an attempt is made to see how these are changing and are likely to change further in the future. The work at this stage falls into two parts; (i) the calculation of average trends in parameters based on simplified assumptions; and (ii) the modification of these preliminary values in the light of discussion with technicians and business men. The first step provides an initial picture derived by communicable processes as a basis for discussion; the second step enables the results of the first to be modified by more subtle processes with a larger content of knowledge and a smaller content of assumption.

The small networks at the bottom of the main diagram show how the first stage is achieved.

Consider first the treatment of the current input–output matrix, A. This is based initially on statistics of the past [174]. The next step is to bring these statistics up to the base-year, 1960. This is done by assuming that the coefficients change over time as a consequence of three factors: (i) changes in prices; (ii) changes in the substitution of one product for another, for example the substitution of electricity for coal; and (iii) changes in the degree of fabrication, that is the value added to materials. If factor (ii) applies with equal force in all industries, for example if electricity is being substituted for coal at the same rate in every industry, then this factor can be represented by the elements of a diagonal matrix which premultiplies the initial A. Similarly, if factor (iii) applies in any industry with equal force to all inputs, then this factor can be represented by the elements of a diagonal matrix which postmultiplies the initial A. Thus

$$A_{60} = \hat{m} - \hat{p} A_{54} \hat{p}^{-1} \hat{n}^*$$ (16)

7

where A_{60} is the base-year matrix, A_{54} is the initial matrix, p is the price vector connecting the 1960 price levels with the 1954 ones, and m^* and n^* are vectors whose elements represent factors (ii) and (iii). These vectors can be computed by an iterative adjustment process described in [156, XVI] if intermediate output and input vectors are calculated for the base-year. This can be done by estimating industry outputs, final products and primary inputs as shown in [156, XVI]. It is shown in [28] that the adjustment process converges to a unique solution for non-negative values in the initial matrix.

Having reached by these means a base-year matrix it is possible to allow for further changes due to factors (ii) and (iii) by deriving new vectors, m and n, from the experience summarized in m^* and n^*. Thus A_0, the coefficient matrix with which we start in some future year, is given by

$$A_0 \, \hat{m} = A_{60} \hat{n} \tag{17}$$

Using (17) we calculate a price vector p_0 from

$$p_0 = (I - \hat{n} A_{60}' \hat{m})^{-1} f_{60} \tag{18}$$

where f_{60} is the vector of factor costs per unit of output. In the course of the main calculations p_0 is changed to p_1, and so at the second round A_0 is replaced by

$$A_1 \, \hat{p} = {}_1 A_0 \hat{p}_1^{-1} \tag{19}$$

The capital matrix, K, is handled in a different way. The starting value, K_0, used in the first round is based on extrapolations of capital–output ratios and of the composition of the stock of assets in each industry in terms of construction, plant and machinery, and vehicles. In the course of the first round a provisional output vector, q_{01}, and a provisional stock vector, k_{01}, are generated. Accordingly, in the second round K_0 is replaced by

$$K_1 = \hat{p}_1 M \hat{p}_1^{-1} \hat{q}_{01}^{-1} \hat{k}_{01} \tag{20}$$

where M is a matrix each column of which shows an industry's proportionate composition of assets. Thus not only p and μ but also A and K change at each round in the calculations.

The methods that have just been described are intended as a means of freeing the model at least to some extent from complete dependence on the observed technology of the past. Of course it would be better to feed in direct knowledge about probable changes instead of making simplified assumptions about them. But technical experts, who are not as a rule expert in building complicated economic models, are likely to be more stimulated and more useful if they are given something to shoot at and can see the framework into which their contribution is to fit. Consequently we see the methods that have just been described as the first step in building a realistic model of the future and as providing agenda for detailed discussions with practical men, after which a true picture should emerge.

8. SOME REALISTIC COMPLICATIONS

As I said in the introduction, this paper is one of a series and I have intentionally avoided overloading it with practical details that have been discussed elsewhere. A number of these will now be considered just to show that they are not incompatible with the present version of the model.

First, consumption has been treated as if there were a single commodity classification equally applicable to producers and consumers. In fact this is not so, but the two classifications can be reconciled by means of a classification converter as exemplified in [XVI]. The demand functions apply to the consumers' classification the elements of which are then transformed into the elements of a producers' classification.

Second, nothing has been said here about final demands for additions to social capital, such as houses, hospitals, schools, roads and government offices. These demands have, of course, to be estimated and can be entered as committed quantities, elements of b^*, in the demand functions.

Third, nothing has been said about foreign trade because the present version of the model adds nothing on this score to the earlier ones. Like requirements for additions to social capital, export demands must be estimated and imports must be such that, given a value for the balance of trade, the external account as a whole is in balance. This balance should be brought about by a price sensitive mechanism such as is introduced here in the consumption and production functions. To do this it would be necessary to have elaborate demand functions for imports and exports and, what is more daunting, foreign shadow prices as well as domestic ones. For the time being therefore, the external account is balanced in a more perfunctory way. Imports are divided into complementary imports which must effectively come from abroad and competitive imports which could, if necessary, be replaced by domestic production. The former are assumed to be mainly intermediate and to be proportional to the output of the using industry, the latter are assumed to be a linear function of the amount of money available from exports after complementary imports, net income from abroad and foreign lending have been accounted for. On these assumptions, imports are in part dependent on production levels and in part a (negative) element of final demand the components of which are calculable in terms of total exports and the balance of trade.

Fourth, net investment demands, v, have been treated as demands for fixed assets but can be increased to allow for stock building proportional to the increase in output even if these stocks are not considered as entering the production functions.

Fifth, replacement demands have been ignored. They are regarded as exogenous, being dependent on the assumed life spans of different assets and the amounts installed around the relevant dates in the past.

Finally, indirect taxes and subsidies have been ignored, but they are not a serious complication provided the rates per unit of output are assumed to be given. If these rates are elements of a vector, t, then p in (15) is equal to a matrix multiplier times $(f+t)$.

9. THE MODEL AND ITS PURPOSE

We may imagine the model set up for a starting date in the future, say 1965 or 1970. The variables that can be set at will are: (i) the utility level, μ; and (ii) the growth vector for consumption, r. If we set these variables so as to perpetuate the *status quo*, that is to say to perpetuate growth in the components of consumption at the rates attained in the 1950's, we may reasonably expect that the model will accommodate them without raising difficulties. A more interesting problem is to set μ so as to reflect a continuation of past trends up to the starting date and to set r so that the trends in the components of consumption become larger from that date on. Since the efficiency factors in the production functions, the a_s of (8), are assumed to grow according to past trends, this raising of our sights for the future is likely to be reflected in a rise in industrial requirements for assets, k.

It may be that the quantity of assets required for the starting date is so large that the projected situation appears unattainable in the time allowed because the transitional period is too short to build up the necessary stocks. This however does not follow.

In the first place, the absorption of technical knowledge and the introduction of new types of equipment connected with it might proceed much more rapidly if business men believed that there would be sufficient demand for their products and consequently called in more technicians to step up productivity. A model, like the present one, which is designed to discover the pattern of requirements on a wide range of assumptions about future demand and growth could help materially in showing how worthwhile it would be to cater for demands in the future far in excess of what could be expected from a projection of past trends.

In the second place, an apparent need to substitute capital for labour on a large scale would be greatly affected if it were found that the present industrial structure contained a significant amount of disguised unemployment. It is an observed fact that the technical processes in Britain and America are often the same but that the number of men behind the processes is much larger here than in America. To the extent that this is true we suffer from disguised unemployment and it is as necessary to convince the trade unions that the jobs would be available if they relaxed their restrictive practices as it is to convince the business world that the demand would be forthcoming if they increased the rate at which new techniques are adopted. The model I am describing should help to provide this demonstration because the range of possibilities it opens up would be greatly increased if a substantial part of the nominal labour force embodied in λ could be released for effective employment.

In my opinion the breaking down of these two fears, over-production leading to low profits and unemployment leading to low wages, would transform the economic life of Britain. The practical purpose of our model is to help in effecting this breakdown by demonstrating that a well-remunerated and profitable situation can exist in conditions quite different from those to which we are accustomed, conditions in fact which would put us nearer the top than the bottom in the scale of economic growth.

I would go further than this and say that a clear indication that our problem of growth can be solved and an assurance that the leading forces in the community are setting about its solution are the necessary conditions for achieving stability in wages, profits and prices. No thinking Briton can believe that the standard of living of this country can only rise by 2 per cent a year. If the present arrangements provide no more than this, who can be blamed for trying strong-arm methods ? That these are futile is obvious but in the meantime we are destroying the advantages in foreign trade, to say nothing of political prestige, that we had inherited from the past. The position *is* more difficult now, but to a large extent these greater difficulties are of our own making.

VIII

TRANSITIONAL PLANNING

Freedom does not consist in the dream of independence from natural laws, but in the knowledge of these laws, and in the possibility this gives of systematically making them work towards definite ends. This holds good in relation both to the laws of external nature and to those which govern the bodily and mental existence of men themselves – two classes of laws which we can separate from each other at most only in thought but not in reality. Freedom of the will therefore means nothing but the capacity to make decisions with knowledge of the subject.'

F. ENGELS, *Anti-Dühring*, Part I, Chapter XI

1. INTRODUCTION

In this paper I shall propose a means of adapting an economic system from one rate of growth to a higher one. I shall take as the indicator of growth, real consumption, private and public, per head of the population. In Britain, this indicator has shown an average peace-time rate of growth of about 2 per cent a year over the last half century. Suppose that, after a suitable transitional period, we wanted to change this rate to $\rho > 2$ per cent; how should we set about it?

The first thing to do is to decide on a value for ρ. Unless we know what growth rate we are aiming at, we can hardly set about the problem of adaptation. If we set ρ too low we condemn ourselves to a needlessly slow increase in our standard of living; if we set it too high we impose on ourselves an uncomfortable and perhaps unattainable burden of inventiveness, adaptability and organization. Somewhere between the too high and the too low we may expect to find a growth rate which most of us would be willing to aim at if we realized what it involved. If we succeeded in achieving it and experienced living with it for a time, we might decide to revise it, up or down. But that is something for the future with which I am not concerned.

The problem of choosing an initial aim, in the form of an initial value of ρ, is the subject of the first numbers in a series of papers entitled *A Programme for Growth* on which a group of us are working at the Department of Applied Economics in Cambridge. I assume therefore that we have settled the question of ρ and that we have also settled two related questions, at least provisionally: first, how long a time do we allow to adapt the economy to achieve a growth rate of ρ; and, second, what level of consumption per head do we assume to have been achieved at the end of the transitional period.

With this preamble I can now define the subject of this paper more exactly; what changes in economic variables should take place during the transitional period if we are to attain our objective, and do we need to revise our preliminary views either as to the length of this period or as to the level of consumption to be reached at the end of it?

This paper is offered to one who throughout my professional life has upheld the torch of econometrics. Being an econometrician myself, I shall try to make my contribution accord with the econometric rules of the game. This means that I shall work with a model which can be given a numerical, empirical content. This is not the place, I feel, to indulge in pure theory which, though it might lead to valuable insights, did not embody the means of reaching quantitative results which the politician, the administrator and, above all, the man in the street have a right to expect of the economist. Insight is good but calculation, even if initially not very precise, is better since it enables us to explore alternative policies in detail without suffering their unwanted consequences in real life.

The model used here is closely related to the one described in the first issue [30] of *A Programme for Growth*. Throughout this paper I shall use a matrix notation and it may help the reader if I set out at this point the conventions on which it is based.

(i) In almost every case a capital letter denotes a matrix; these are in all cases square and of an order equal to the number of groups of products distinguished in the model. The exceptions, mainly Greek capitals, are the operators: Σ for summation; Π for forming a product; E for shifting a variable (thus $E^\theta \mu(\tau) \equiv \mu(\tau+\theta)$); and Δ for forming first differences (thus $\Delta \equiv E-1$).

(ii) Small Roman letters denote vectors. These are written as column vectors: a row vector is written with a prime superscript, as is the transpose of a matrix. The letter i is always used for the unit vector: that is $i \equiv \{1, 1, ..., 1\}$ where $\{\}$ denotes that the elements of a column vector are written out in a row. With one exception, diagonal matrices are denoted by the symbol for a vector surmounted by a circumflex accent. The exception is the familiar I, used in place of $\hat{i}$ to denote the unit matrix.

(iii) Small Greek letters denote scalars with the exception of ϕ which denotes a function.

With these conventions the nature of the different symbols soon becomes apparent, and scalar and matrix algebra can be mixed without confusion or the need for a bold and forbidding type face in which to represent matrix symbols.

2. ASSUMPTIONS

For present purposes the characteristics of the model can be set out as follows.

First, at the outset of the transitional period, the vector, s, of the economy's stock of capital goods is equal to a given value, $\bar{s}$. That is

$$s \equiv \bar{s} \tag{1}$$

The elements of s and $\bar{s}$ relate to the total stock of different capital goods, grouped according to the producing industry, not according to the using industry.

Second, at the end of the transitional period, which runs from $\theta = 0$ through $\theta = \tau - 1$, the terminal stock vector, $E^\tau s$, must have a certain minimum value, $\bar{s}$, to provide for the direct and indirect requirements of the level of consumption provisionally assumed for year τ and of the investment in that year needed to achieve the assumed rate of growth from that year on. That is

$$E^\tau s \geqslant \bar{s} \tag{2}$$

Third, at all times there must be adequate capacity to make possible the level of production decided on. If q denotes the output vector and K the capital input–output coefficient matrix, then

$$E^\theta s \geqslant KE^\theta q \tag{3}$$

Fourth, at all times the labour force must be large enough to make possible the level of production decided on. If λ denotes the total labour force and f the vector of labour requirements per unit of output in the different industries, then

$$E^\theta \lambda \geqslant E^\theta f' q \tag{4}$$

Fifth, the usual input–output flow equation can be written as

$$E^\theta q = AE^\theta q + E^\theta e + E^\theta v$$
$$= (I-A)^{-1} E^\theta (e+Es-s) \tag{5}$$

where e and v denote respectively the vectors of consumption and investment and A denotes the current input–output coefficient matrix. In the second row of (5) I have made use of the equality $v = \Delta s$ which implies that I ignore for the moment investment for replacement purposes and concentrate on extensions. I shall return to this point in section 6 below.

Sixth, the demand equations are of the linear expenditure form and can be written as

$$\hat{p}e = b\mu + (I - bi')\hat{c}p$$
$$= \hat{p}c + b(\mu - p'c) \tag{6}$$

where μ denotes total expenditure, p denotes the price vector and e denotes a vector of quantities bought. If there are ν products then the vectors b and c together contain $2\nu - 1$ independent constants since $i'b = 1$. They can be interpreted as follows: the average consumer buys amounts of the different products equal to the elements of c at a total cost of $p'c$; he then allocates the remaining money he has to spend, $\mu - p'c$, to the different products in proportion to the elements of b.

This system of demand equations is designed primarily to represent private consumers' behaviour. It is quite possible that it could represent public consumption as well. Alternatively, public consumption could be taken as given year by year. This second treatment would certainly be necessary for government expenditure on social capital: schools, hospitals, roads and the like.

In this discussion of demand I have said nothing about foreign trade. I shall return to this important subject in section 6 below.

3. CONSTRAINTS

During the transitional period we should like to enjoy as high a standard of living as possible. However, during this period the requisite terminal stocks given in (2) must be built up. Consumption through the transitional period is therefore constrained in various ways.

The first kind of constraint comes from the actual level of initial stocks and the required level of terminal stocks. These constraints are expressed in (1) and (2) above.

The second kind of constraint comes from the fact that in each year we must have enough labour and capital goods to produce the output of that year. For labour, this constraint is expressed in (4); for capital goods it works out as follows.

By premultiplying (5) by K it follows that

$$KE^\theta q = K(I-A)^{-1} E^\theta(e+Es-s)$$
$$= FE^\theta(e+Es-s) \tag{7}$$

where $F \equiv K(I-A)^{-1}$. By combining (3) and (7) we see that

$$E^\theta s \geqslant FE^\theta(e+Es-s) \tag{8}$$

or that

$$FE^\theta e \leqslant (I+F)E^\theta s - FE^{\theta+1} s \tag{9}$$

which, when written out in full for the whole of the transitional period, gives

$$\begin{bmatrix} Fe \\ FEe \\ \vdots \\ FE^{\tau-1}e \end{bmatrix} \leqslant \begin{bmatrix} (I+F) & -F & \dots & 0 & 0 \\ 0 & (I+F) & \dots & 0 & 0 \\ \vdots & \vdots & & \vdots & \vdots \\ 0 & 0 & \dots (I+F) & -F \end{bmatrix} \begin{bmatrix} s \\ Es \\ \vdots \\ E^\tau s \end{bmatrix} \tag{10}$$

At this stage I am assuming that A, K and f are constant. If we work in terms of a given wage rate and assume that the return on capital bears a constant ratio to it, then it follows that p is also constant.

By premultiplying (6) by $F\hat{p}^{-1}$ it follows that

$$FE^\theta e = Fg + FhE^\theta \mu \tag{11}$$

where $g \equiv (I - hp')c$ and $h \equiv \hat{p}^{-1}b$. When written out for the whole transitional period this gives

$$
\begin{bmatrix} Fe \\ FEe \\ \vdots \\ FE^{\tau-1}e \end{bmatrix} = \begin{bmatrix} Fg \\ Fg \\ \vdots \\ Fg \end{bmatrix} + \begin{bmatrix} Fh & 0 & \dots & 0 \\ 0 & Fh & \dots & 0 \\ \vdots & \vdots & & \vdots \\ 0 & 0 & \dots & Fh \end{bmatrix} \begin{bmatrix} \mu \\ E\mu \\ \vdots \\ E^{\tau-1}\mu \end{bmatrix}
\tag{12}
$$

Finally, by combining (9) and (11) we obtain

$$
FhE^{\theta}\mu \leqslant (I+F)E^{\theta}s - FE^{\theta+1}s - Fg
\tag{13}
$$

Equation (9) states that the capital goods needed directly or indirectly to produce consumption goods in year θ cannot exceed the amount available at the beginning of that year less the amount needed directly or indirectly to produce capital goods in that year. Equation (11) gives the linear form of the Engel curves for consumption goods implicit in (6). Equation (13) results from combining the two.

A third kind of constraint comes from the fact that we should probably wish to ensure a minimum level of consumption in each year of the transitional period. This can be expressed as

$$
E^{\theta}\mu \geqslant \mu^{*}
\tag{14}
$$

where μ^{*} is a preassigned minimum level. Alternatively we might prefer to ensure that consumption did not fall during the transitional period. This can be expressed as

$$
E^{\theta}\mu \geqslant E^{\theta-1}\mu
\tag{15}
$$

Subject to the constraints (1), (2), (4), (13) and (14) or (15) we can do what we like. The question now is: what do we like?

4. A MAXIMAND

In principle we should like to maximize the total utility of consumption over the transitional period. If we use the subscript $\zeta, \zeta = 1, 2, \dots, \nu$, to denote the typical commodity, then the utility of consumption in year θ, $E^{\theta}v$ say, is, with demand functions of the form of (6),

$$
E^{\theta}v = \phi\left[\prod_{\zeta=1}^{\nu} (E^{\theta}e_{\zeta} - c_{\zeta})^{b_{\zeta}} \right]
\tag{16}
$$

where ϕ denotes an arbitrary monotonic function. Accordingly, over the whole transitional period, the total utility of consumption is

$$
\sum_{\theta=0}^{\tau-1} E^{\theta}v = \sum_{\theta=0}^{\tau-1} \left\{ \phi\left[\prod_{\zeta=1}^{\nu} (E^{\theta}e_{\zeta} - c_{\zeta})^{b_{\zeta}} \right] \right\}
\tag{17}
$$

This is not a very promising expression: first, because of the arbitrary nature of ϕ; and, second, because of the way in which the $E^\theta e_\zeta$ enter into the determination of total utility.

In practice therefore we must find a more tractable maximand. Since prices are assumed to be constant, it seems reasonable to aim at giving the average consumer the maximum amount of money to spend on consumption over the whole transitional period and leave him to spend it to his own best advantage. If we accept this line of thought we simply maximize

$$\sum_{\theta=0}^{\tau-1} E^\theta \mu$$

5. A LINEAR PROGRAMME

The preceding two sections lead to the following linear programme.

$$\sum_{\theta=0}^{\tau-1} E^\theta \mu = \max \tag{18a}$$

subject to

$$s = \bar{s} \tag{18b}$$

$$E^\tau s \geqslant \bar{\bar{s}} \tag{18c}$$

$$E^\theta \lambda \geqslant E^\theta f' q \tag{18d}$$

$$FhE^\theta \mu \leqslant (I+F) E^\theta s - FE^{\theta+1} s - Fg \tag{18e}$$

and using (14) for the sake of argument,

$$E^\theta \mu \geqslant \mu^* \tag{18f}$$

The attempt to solve this problem may show that there is no acceptable solution. Formally, we can always say that this is because μ^* has been set too high, but the real reason is likely to be that $\bar{s}$ has been set too high. This in turn is either because the level of consumption at the close of the transitional period, $E^\tau \mu$, has been set too high, or because the rate of growth in consumption after the transitional period, and therefore the investment required initially to sustain it, has been set too high.

By assumption, the ultimate rate of growth has been fixed by the long-term calculations and so if there is no acceptable solution to the original problem, the next step is to reduce $E^\tau \mu$. The effect of this will be to increase the average level of $E^\theta \mu$, $\theta = 0, 1, \ldots, \tau - 1$. It may still happen, however, that there is a big jump between $E^{\tau-1} \mu$ and $E^\tau \mu$. This would signify that the terminal level of consumption was excessive in relation to the transitional level. One remedy would be to reduce the terminal level of consumption still further until there was an acceptable relationship between $E^{\tau-1} \mu$ and $E^\tau \mu$. An alternative remedy might be to lengthen the transitional period, but it is perhaps more interesting to consider this possibility when we relax the assumption of a one-year production period for all forms of capital good.

Thus, in principle, the method of handling the transitional problem proposed here is to solve the linear programming problem (18) and to vary the terminal stock vector $\bar{s}$ until an acceptable relationship is found between $E^{\tau-1}\mu$ and $E^{\tau}\mu$.

In practice, the size of the problem means that we must use dynamic programming methods so that we can build up a solution step by step, that is year by year. I shall not go into computational questions here.

6. SOME RELAXATIONS

In order to present the main solution as simply as possible, I have carried out the argument so far on simplified assumptions which are clearly unrealistic. I shall now try to show how these can be removed, treating the different problems roughly in the order of their difficulty.

(a) *The parameters, b and c, of the consumption functions.* If we consider the vectors b and c in (6) above and the meaning that was there attached to them, we can see that they are most unlikely to remain constant over time. As time goes on and people experience a rising standard of living, they feel committed to more and more and naturally alter the allocation of the money they have available for free spending. The extent to which this has been happening in Britain is demonstrated in [158] and, in more detail, in [160].

The theoretical properties of (6) which include the Slutsky condition of a symmetric substitution matrix would be destroyed by the addition of time trends to the individual demand equations but are not affected if b and c are made functions of predetermined variables. A simple possibility is to make them linear functions of time, and it is shown in [160] that this modification works well and that a slightly more complicated but equally practicable modification would work even better.

Thus the changes in the elements of b and c can be worked out from past experience and projected into the future. All that happens to our linear programme, (18), is that g and h take different known values in each year of the transitional period.

(b) *The replacement of capital goods.* I mentioned in connection with (5) in section 2 above that I would introduce later the question of the replacement of capital goods. On the assumption that these are replaced at the end of a fixed span of useful life, different of course for different capital goods, then a vector of known constants, which change from year to year, must be included in the second bracket of the second row of (5). A more interesting treatment would be to allow for replacement as soon as a plant, working at capacity, fails to earn in excess of the labour cost of operating it. I shall not discuss this extension here, but a model of production which will eventually enable us to give effect to it is described in [129].

(c) *The matrix, A, of current input–output coefficients.* The elements of the matrix A are likely to be estimated initially from past data: for Britain the most recent input–output matrix estimated directly relates to 1954 [174]. These past estimates must be brought up to date and projected into the future. As outlined in [158] and demonstrated fully in [32], this can be done by combining a mathematical adjustment procedure with a limited amount of outside information. The essence of this method can be described as follows. Suppose we have an initial matrix, A_0 say. If p_{10} is a price vector whose elements are the ratio of prices in period 1 to prices in period 0, then we can form $A_0^* \equiv \hat{p}_{10} A_0 \hat{p}_{10}^{-1}$ which expresses A_0 in terms of the prices of period 1. We can then estimate A_1, the matrix appropriate to year 1, as

$$A_1 = \hat{j} A_0^* \hat{k} \tag{19}$$

where j and k are vectors of parameters which can be determined up to a scalar, that multiplies all the elements of j and divides all the elements of k, from a knowledge of A_0, the output vector q_1 and control totals of intermediate input and intermediate output in year 1. Each element of j multiplies the corresponding row of A_0^* and can be interpreted as a substitution multiplier; each element of k multiplies the corresponding column of A_0^* and can be interpreted as a fabrication multiplier.

Experience shows [158] that this method, despite its simplicity, works well for most elements of A with a small number of specific exceptions. A typical example is as follows. In the last decade there was a tendency to replace coal as a fuel by electricity and oil; hence the input of coal per unit of output in different branches of production tended to fall. But in coke ovens coal is a raw material and, naturally enough, the coefficient of coal into coke ovens did not follow the same trend as coal into other uses: in fact it did not fall at all. Fortunately there are up to date statistics of coal used in coke ovens, and so this, and a number of similar abnormal elements, could be removed from A_0, q_1 and the control totals and could then be added back after the remaining elements had been estimated.

The estimate of A_1 can be projected to a future year, to give A_2 say, by using the relationship

$$A_2 = \hat{j}^\theta A_0^* \hat{k}^\theta \tag{20}$$

where the exponent θ is the ratio of the time spans separating years 0 and 2 and years 0 and 1. The results though probably in the right direction may easily become exaggerated over a ten-year period and genuinely new technical developments will only be reflected by chance. Accordingly in making these projections it is important to draw as heavily as possible on the experience and expectations of technicians in the various branches of industry on probable changes in technology.

However the projections of A are made, and in centrally planned economies it may be possible to make them on a firmer basis than is at present available to us in Britain, the immediate effect on the linear programme, (18), is small. All that happens is that F is replaced by $E^\theta F$, different but known for each year. A change in A may also be expected to change p; but I shall come to that under (h) below.

(d) *The matrix, K, of capital input–output coefficients.* Like those of A, the elements of K can be thought of as changing under the influence of relative prices and of row and column multipliers. In this case, the row multipliers reflect the substitution of one capital good for another: in Britain the stock of plant and machinery has tended to grow faster than the stock of buildings in which they are kept. The column multipliers reflect a change in the productivity of the complex of capital goods used in each of the various industries. In the capital analogue of (19) we may expect that the row multipliers for plant and machinery will usually be relatively large and that those for buildings and construction will be relatively small. At the same time, since the productivity of capital goods tends to rise in almost all industries, we may expect that most of the column multipliers will be relatively small.

In principle the method I have just described for the matrix A could be used to bring up to date and to project the capital input–output matrix K. If we postmultiply this matrix by a diagonal matrix formed from the first difference, Δq, of the output vector, we obtain a flow matrix whose row sums are the investment demands on the various industries producing capital goods and whose column sums are the expenditures on capital goods by each of the using industries. These two vectors of marginal totals could be compared with investment statistics but the correspondence would not have the same significance as with the current matrix because the investment in a year may not be very closely related to the change in output in the following year. However, the correspondence will tend to improve if the period is lengthened, particularly if adjustments can be made for working under capacity in certain industries.

In practice it may be better to approach the projection of K by looking at the changes in: (i) the proportion of different kinds of capital good in the stock of assets in the different industries; and (ii) the productivity of the collection of capital goods employed in the different industries. These are both matters on which information is available in Britain. In handling these questions, considerable importance attaches to the measure of capital adopted, a question which has recently been examined by one of my colleagues in [128, 129].

As with A, however the projections of K are made, the immediate effect on the linear programme (18) is simply to cause F to change in a known way from year to year.

(e) *Differences in investment lags.* So far I have assumed that investment undertaken in one year would result in usable capital goods at the beginning of the following year. In general it would not, because many capital goods take more than a year to build and instal. This fact is particularly important in analysing transient states of the economy, though it would be irrelevant if all elements of the output vector were in a state of linear growth. It can be represented by partitioning the capital input–output matrix, K, to show the contribution of different stages to capital goods requiring different lengths of time to build. As described in [30], let us consider the case in which the longest production period of capital goods is three years and define a matrix K^* as

$$K^* \equiv \begin{bmatrix} K_{11} & K_{12} & K_{13} \\ K_{21} & K_{22} & 0 \\ K_{31} & 0 & 0 \end{bmatrix} \tag{21}$$

connected to K by the relationship

$$K = [I \, I \, I] K^* \{I \, I \, I\} \tag{22}$$

The submatrices of K^* are partitions of K. The rows of K^* relate to the stage of the work, the columns of K^* relate to the date of completion and the secondary diagonals of K^* relate to capital goods with different production periods. Thus K_{11}, K_{12} and K_{13} relate to first year's work on capital goods that take respectively one, two and three years to complete; K_{11}, K_{21} and K_{31} relate respectively to first, second and third year's work on capital goods that will be completed at the end of this year; and K_{13}, K_{22} and K_{31} relate respectively to first, second and third year's work on capital goods that take three years to complete.

An economy working at capacity which wishes to achieve certain increases in output in the future must: (i) finish the capital goods needed for the increase in output next year; (ii) do first or second year's work on capital goods needed to increase capacity in the following year; and (iii) do first year's work on capital goods needed to increase capacity in the year after that. Thus

$$v = \Delta s$$
$$= [I \, I \, I] K^* \{\Delta q \, E\Delta q \, E^2 \Delta q\} \tag{23}$$

In general, this investment does not all add to capacity next year but in part results in unfinished capital goods: work in progress. Accordingly s is no longer the concept of stocks which limits production possibilities. Let us therefore introduce a concept of completed stocks, s^* say, where

$$s^* = s - [(K_{12} + K_{13} + K_{22}) + K_{13} E] \Delta q \tag{24}$$

In the course of this year, work will be done to complete the capital goods needed for next year's increase in output. But the earlier stages of the work must have been done in the past if the desired increase is to take place. In this case

$$v^* = \Delta s^*$$

$$= K\Delta q \tag{25}$$

and v is connected to v^* by the relationship

$$v = v^* + [(K_{12} + K_{13} + K_{22}) + K_{13}E]\Delta^2 q \tag{26}$$

If the components of q grow linearly, then $E^\theta \Delta^2 q = \{0, 0, \ldots, 0\}$ and so $v = v^*$ but in general this is not the case and we may expect that $v > v^*$.

Accordingly, to return to the linear programme, (18), we see that we must replace s by s^* and at the same time retain $v = \Delta s$ as the relevant concept of investment. With this modification (10) is replaced by

$$
\begin{bmatrix} Fe \\ FEe \\ \vdots \\ FE^{\tau-1}e \end{bmatrix}
\leqslant
\begin{bmatrix}
I & 0 \ldots 0 & -F & 0 \ldots & 0 \\
0 & I \ldots 0 & 0 & -F \ldots & 0 \\
\vdots & \vdots \; \vdots \; \vdots & \vdots & \vdots & \vdots \\
0 & 0 \ldots I & 0 & 0 \ldots & -F
\end{bmatrix}
\begin{bmatrix}
s^* \\ Es^* \\ \vdots \\ E^{\tau-1}s^* \\ \Delta s \\ E\Delta s \\ \vdots \\ E^{\tau-1}\Delta s
\end{bmatrix}
\tag{27}
$$

This equation states that the capital goods required directly or indirectly to produce the consumption goods demanded in any period must not exceed the stock of completed capital goods available at the beginning of the period less the capital goods required directly or indirectly to complete capital goods and prepare the work in progress needed for future increases in production.

The change from (10) to (27) is not the end of the story: we must recognize that new constraints are imposed on the programme. These arise because at the beginning of the transitional planning period, work in progress is inherited from the past and all that can be done is to complete it: some of the elements of $E^\theta v$ for small values of θ are fixed. It is mainly for this reason that in our work in Cambridge we have taken 1970 as the end of the transitional period rather than, say, 1965. The higher steady growth rate we should like to achieve almost certainly involves an adaptation of the capital structure of the economy which might be achieved in seven years but could not be achieved in two.

As a gloss on this section, it should be added that time lags may arise in the current input–output relationships. The implicit assumption here is that intermediate product can always be produced in the year in which it is wanted. In this case, however, the time lags are certainly shorter than for capital goods and may reasonably be ignored, at least in a first approximation.

(*f*) *Changes in the productivity of labour, f.* Changes in technique, usually embodied in new and better capital goods, alter the labour required per unit of output. This could be allowed for by projecting past trends in labour required per unit of output in different industries, thus replacing the original fixed f by $E^{\theta}f$ which changes in a known way with time.

This treatment of f, like the corresponding treatment of K in (*d*) above is unsatisfactory in two ways. First, it is to be expected that the achievement of a higher rate of growth will call for the introduction of new techniques at a rate faster than that observed in the past. Accordingly, we shall probably need to experiment with trends in the elements of f and K larger than those observed in the past. This, of course, is only a beginning: in itself, it is mere wishful thinking. The next step is to see whether there are compelling obstacles to the higher rates assumed. Perhaps there are; but perhaps the real obstacles are a fear of oversupplying the market on the part of management or of unemployment on the part of labour. But these fears may be based on an assumed continuation of the slow rate of growth of output observed in the past and their rational basis would be largely removed if it were shown that this rate could be increased substantially.

Second, the relationship of value added to primary inputs at present embodied in the model is of a very simple kind: changing coefficients relating capital goods and labour to net output. This treatment must be abandoned before long in favour of one which shows how far capital and labour can be substituted for one another in different industries. Only in this way can we discuss the question of the optimum allocation of labour and derive shadow prices that are demonstrably related to efficiency. Before abandoning the simple primary input–output ratios, however, we want to make sure that the production functions we have in mind do in fact accord with past observations. These production functions are based on the idea that there are limits at any one time to primary input proportions and that the possibilities of substitution are largely confined to new plant and equipment [129].

(*g*) *Foreign trade.* So far I have assumed a closed economy, but foreign trade is so obviously important that a solution without it would be no solution at all. As explained in [30], there are several ways in which imports and exports can be introduced into our long-term model. All fall short of the ideal largely because even if we can calculate domestic shadow prices for the future, we cannot do the same for all the countries with which we trade. Accordingly the details of our foreign trade cannot be determined by considerations of relative price and we must fall back therefore on a simplification.

The simplest way of introducing foreign trade can be described as follows. Let us begin by rewriting (5) as

$$E^{\theta}q = AE^{\theta}q + E^{\theta}e + E^{\theta}v + E^{\theta}x - E^{\theta}m_1 \tag{28}$$

8

where x denotes a vector of exports and m_1 denotes a vector of competitive imports, that is imports of products which directly compete with a similar domestic product. Let us next define the balance of trade, β say, as

$$E^\theta \beta = E^\theta(p' x - p_1^{*'} m_1 - p_2^{*'} m_2) \tag{29}$$

where m_2 denotes a vector of complementary imports, that is imports of products either not made domestically or made in only small quantities; and where p_1^* and p_2^* denote respectively the price vectors of competitive and complementary imports. If we want a component of m_2 we must buy it from abroad and the amount we want depends on the output of different industries. In other words

$$E^\theta m_2 = \hat{a} E^\theta q \tag{30}$$

If $E^\theta \beta$ is fixed then $E^\theta p_1^{*'} m_1$ is the residual element in the balance of payments. Let us suppose that the elements of m_1 are in part fixed and in part responsive to the amount of money available for competitive imports, $E^\theta p_1^{*'} m_1$. Then we might put

$$E^\theta \hat{p}_1^* m_1 = E\theta \hat{p}_1^* a^* + a^{**}(E^\theta p_1^{*'} m_1) \tag{31}$$

In (30) and (31), a, a^* and a^{**} are parameters based on past experience. I shall not consider the complication that they may all be changing over time since if we knew how they were changing the present argument would be unaffected.

Equations (29) through (31) enable us to rewrite (28) as

$$E^\theta q = (I - A - \hat{p}_1^{*-1} a^{**} a' \hat{p}_2^*)^{-1} E^\theta[v + e + (I - \hat{p}_1^{*-1} a^{**} p') x + \hat{p}_1^{*-1} a^{**} \beta - a^*] \tag{32}$$

Thus we see that (5) remains as before except that: A is replaced by $A + \hat{p}_1^{*-1} a^{**} a' \hat{p}_2^*$; and the term $E^\theta(v + e)$ is augmented by a term in exports, $E^\theta(I - \hat{p}_1^{*-1} a^{**} p') x$, and a constant term, $E^\theta(\hat{p}_1^{*-1} a^{**} \beta - a^*)$. With these modifications we can proceed as before.

With these changes, the solution of (18) will ensure that output is sufficient to meet export demands and that the intended balance of payments in each year is realized. But we may not like the results because competitive imports may be very considerably squeezed. To remedy this we must either plan to export more, find substitutes for complementary imports or content ourselves with a less favourable balance of payments. Any combination of measures to these ends will help to improve the position of competitive imports.

(h) *Future shadow prices.* Most of the modifications of the model that have been introduced in this section not only lead directly to changing coefficients but also indirectly to changing prices. A change in p has two effects which we must now consider: first, it changes g and h in (18e); and second, it changes the maximand in (18a).

In order to see how to calculate the $E^\theta p$, let us begin with the simplest case in which there is only one primary input whose rate of reward, ω say, is the *numéraire* of the whole system. As before, let the suffixes 1 and 2 denote respectively the base year and any of the years in the transitional period. For the time being I shall put $\omega_1 = \omega_2 = 1$.

The units of quantity in the model are in each case the £'s worth at base-year values. Accordingly, denoting by y a vector whose elements represent the quantity of primary input per unit of output in each of the industries in the base year, we can write the price equation for that year as

$$p_1 = A_1' p_1 + y_1$$
$$= (I - A_1')^{-1} y_1$$
$$= i \tag{33}$$

since

$$y_1 = (I - A_1') i \tag{34}$$

In relation to the prices in year 1, prices in year 2 are the elements of a vector p_{12} say, where

$$p_{12} = (I - A_2^{*\prime})^{-1} \hat{r}_{12} y_1 \tag{35}$$

In this equation $A_2^{*\prime}$ is the transpose of the input–output coefficient matrix in year 2, the elements being valued at the prices of year 1, and the elements of r_{12} are the reciprocals of the changes in productivity of primary inputs in the various branches of production. Thus p_{12} can be measured from our projections of input–output relationships and of trends in the productivity of primary inputs.

It is perhaps of interest to note that $A_2^{*\prime}$ is connected to A_2', the corresponding matrix recalculated at the prices of period 2, by the relationship

$$A_2^{*\prime} = \hat{p}_{12} A_2' \hat{p}_{12}^{-1} \tag{36}$$

This equation when combined with (35) gives

$$\hat{r}_{12} y_1 = \hat{p}_{12} (I - A_2') i \tag{37}$$

In (37), the elements of $(I - A_2') i$ are the proportions of the price of each product which accrues to primary inputs. When each proportion is multiplied by the corresponding price of the product we obtain, as we should expect, the value of primary inputs per unit of output.

The next step is to divide primary inputs between labour and capital goods. Here a new problem arises. We may interpret ω as the wage rate but now we must decide how, in the future, the return on capital, α say, will be related to it. In order to answer this question properly we need production functions; using coefficients we can only base the connection on a projection of past relationships.

Finally, allowance should be made for complementary imports into production which, together with labour and capital goods, enter into productive processes from sources other than current domestic production. Here we are likely to want to make realistic projections of price trends and so in the end we are led to making projections of ω and of the relationship of α to ω.

If we do all these things, we can estimate future shadow prices, $E^\theta p$. These estimates could be improved if we could substitute production functions for primary input coefficients and trading functions for the simplified treatment of foreign trade described above.

This brings me to the second change required by the assumption that future prices are different from base-year prices: the maximand must be adjusted to a constant-utility basis. As shown by Klein and Rubin [85] the ordinal utility function (16) is associated with a constant utility price index-number of the form

$$\pi_2 = [p_2' c + (\mu_1 - p_1' c) \gamma_b]/\mu_1 \tag{38}$$

The numerator of (38) consists of two terms: the cost in year 2 of the fixed purchases, the elements of c; and the value of free expenditure in year 1 changed by a geometric index of prices, γ_b, with the elements of b as weights. Thus if we replace the maximand in (18a) by

$$\sum_{\theta=0}^{\tau-1} [E^\theta(\mu/\pi)]$$

we shall effect the necessary adjustment. If b and c vary as described in (*a*) above we have a choice of the year whose utility level is to be taken as a base.

Under headings (*a*) through (*h*) of this section I have tried to remove the main over-simplifications to which the linear programme of section 5 is subject, and to do so in such a way that each modification is computable with the data at our disposal. From a theoretical point of view the weakest link is perhaps the connection used between net output and primary inputs and, as I have said, we hope before long to improve this part of the model.

7. CONCLUSION

By way of conclusion I shall indicate my conception of the practical role of the kind of model I have described. It is in the first place to demonstrate the possibilities of a conscious adaptation of the economy and to provide agenda on the main problems of doing so; and in the second place to provide a moving point of reference around which political and administrative measures can be organized and which can serve as a basis for individual economic decisions by labour and management in different industries. It can help to hold the balance between the need for centralized decisions if economic adaptation is to be undertaken consciously and the need to decentralize an enormous number of decisions if full use is to be made of individual knowledge and initiative and if the central administration is not to become hopelessly overloaded and consequently somewhat rigid and inefficient. I believe that computable models will come to occupy an essential place in economic adaptation. But the possibility of using them rests on their acceptance by society; it is men who change the course of events, though they do so through ideas.

Being an optimist, I believe there are signs that this general point of view is gaining ground in many countries, socialist and capitalist alike. Perhaps in the next generation we shall see some big strides forward in economic organization made possible by developments in political thinking, in economic and statistical science, to which the recipient of this essay has contributed so much, and by the emergence of the computer as a new force in social administration.

IX

A MODEL OF THE EDUCATIONAL SYSTEM

1. INTRODUCTION

The purpose of this paper is to outline a model of the educational system designed to work out the present implications of future levels of educational activity as determined by the evolution of the demand for places on the one hand and the economic demand for the products of education on the other. The need for such a model must be apparent to all who have read the report of the Robbins Committee [177] and considered the enormous organizational problems that it sets. It is equally apparent if we consider the possibilities for spectacular improvement which applied science offers to the techniques of production. Great changes are essential if we want to achieve a substantial increase in the growth rate of our standard of living in Great Britain, along the lines that the National Economic Development Council is aiming at, and if Great Britain is to present a less flabby economic image on the international scene.

In terms of *A Programme for Growth* [30 to 36] with which the present work is closely linked, the model described here is simply an example of a sub-model of a complex and economically important process. In trying to model economic and social activity it seems better to think in terms of a set of linked models operating by means of the exchange of information rather than in terms of a single monolithic structure. It is for this reason that I am planning to elaborate the educational model separately from the purely economic model, though I hope that eventually it will be possible to get the two models to interact.

The arrangement of the paper is as follows. I begin by defining the scope of the model to include all forms of education, training and retraining and regard the educational system thus defined as a system of connected processes exactly as in input–output analysis. As far as students are concerned these processes are not interdependent but form a one-way hierarchy of dependence, the lower processes always feeding the higher ones. Each process must provide each year for the graduate leavers (if any) at the end of the year plus the student inputs required from the process for the next year. Thus the flow equation of the system is as in an open, dynamic input–output model and so this year's activity levels expressed in terms of students can be written as a convergent series whose terms are functions of future vectors of graduate leavers.

A peculiarity of this productive system is that it is supplied with an exogenous supply of raw material in the form of young children entering its lowest stages. These can be called the human inputs, as opposed to the economic inputs. Thus, initially, activity levels are largely determined by the demographic characteristics of the population which are represented in the model by a simple birth and survival process. But, beyond a certain point, the decisions of students and their advisers become the main determinant. These are all-important after the age of compulsory education has been passed and give rise to the demand for places. These demands are represented in the model as a multi-stage epidemic. At each stage, the number who catch the infection and decide to go to a university depends partly on the numbers who have gone and so may be presumed to spread the infection, and partly on the numbers who have not gone and so are available to be infected. A number of parallel diseases, corresponding to different specializations, is allowed for at each stage. The total number at each stage is thought of as determined by educational ambitions, the composition of this number being strongly influenced by the economic prospects of the different specializations. These demand influences determine the future mixes of graduate leavers which, as already described, determine current activity levels in the different processes.

The purpose of calculating these activity levels is to enable us to calculate the requirements for economic inputs: teachers, buildings, equipment and supplies. Although supplies (intermediate economic inputs) are not likely to give much trouble, investment programmes, needed to provide new buildings and equipment, usually take some time to complete and have to be considered along with competing claims on the output of the industries producing investment goods. Teachers provide the most difficult problem of all, a problem considered here in terms of the factors determining the supply of and demand for teachers and the influence of the status and pay of teachers in affecting a balance.

As in other branches of the economy in which increased output is required, it can be expected that technical change will also be important in education. I show that any anticipated changes in educational technology, whether they affect learning times or economic inputs, can easily be introduced into the model. Finally, I suggest that the probable importance of radical technical changes provides one reason why the comparatively simple, demand-determined input–output model suggested will eventually have to be reformulated as a model of the programming type. But before this is done it will be useful to look at the data needed to make this simpler model work.

While developing these ideas, I have had the opportunity of discussion and correspondence with my friends Claus Moser and Philip Redfern, whose plans for work in this area are indicated in recent papers [110, 111]. At this stage our ideas are similar but not identical. I am also indebted to Kenneth Wigley who is working with me at the Department of Applied Economics on the construction of economic sub-models. But I alone am responsible for the inevitable shortcomings of this *ballon d'essai*.

2. THE STRUCTURE OF EDUCATIONAL PROCESSES

Before we consider how students enter and leave the educational system, we must try to visualize this system in a way that is useful for analysis.

We can regard the educational system as a set of processes through which students pass from early childhood until they graduate out of the system altogether from one of the processes. In economic terms we might say that students enter the system as raw material, trickle through it as increasingly fabricated work in progress and pass out of it as final product.

Just as in social accounting we have to decide what activities to include in the concept of production, so in the present case we must decide what activities to include in the concept of education. In the first case we use the idea of a production boundary which is notionally drawn round the activities to be included. Not all countries draw this boundary in the same place. The most restrictive boundary is used by the socialist countries which employ the material products concept. This boundary is drawn round activities concerned with the production and handling of goods and excludes a number of service activities such as government services and passenger transport. The boundary used in the system of national accounts drawn up by the United Nations includes these other service activities but excludes the unpaid services of household members and amateurs. The reasons for drawing the boundary in any particular place, and the conventions and imputations needed to obtain a manageable system, need not concern us here. The point is that a precise boundary must be drawn; it does not exist in nature.

A similar problem arises in defining the educational system. As with the economy at large, it is convenient to draw the educational boundary so as to exclude the educative effects of family members on one another; these may be regarded as a part of life in general for which we do not try to account. We are left, then, with various institutionalized forms of education, mainly taking place in schools, colleges and universities. As in the case of social accounting, family activities are excluded not because they are unimportant but because they are not the subject of general policy decisions and because they are virtually unrecorded.

The identification of the educational system with the activities of schools, colleges and universities is not, however, satisfactory: in one way it is, perhaps, too broad; in another way it is too narrow. If we begin at the younger end of the age distribution we find nursery schools and kindergarten whose function is in large part to release the mother from the constant minding of small children. These activities, like family activities, are important, but for similar reasons it does not seem necessary in the first instance to introduce them into an educational model, though eventually they will be needed in connection with the demand for teachers.

At the older end of the age distribution students may become members of industrial firms or professional bodies which provide training for the qualifications needed to practise specific skills. After apprenticeship, an individual may be recognized as a qualified carpenter or fitter; after passing a professional examination he may be recognized as a qualified lawyer or accountant. Since one of the purposes of the model is to link educational qualifications with the skills required in the economy, it is desirable that all these forms of further training and retraining should be included among the educational processes.

In the light of such considerations as these, we can define the educational system by drawing a boundary round a number of processes or activities carried on not only in schools, colleges and universities but also in other institutions and establishments which provide technical or professional training.

The flow of students in this system can be represented by an input–output accounting matrix [151] in which the entries relate to numbers of students rather than to sums of money. Provision must be made for primary inputs (mainly young children entering their first year at school) and for final outputs (graduates leaving the whole system from one of the processes). Apart from this, each row and column pair relates to a process, the outputs (intermediate and final) being shown in the row and the inputs (intermediate and primary) being shown in the column.

In order to see the kind of input–output model that emerges from this way of looking at the educational system, let us consider a simple case in which during the years of education a student enters the system at a given age and passes to a new process each year until at some stage he leaves the system altogether. Let us represent the stock of students at the beginning of a year by a vector, s, whose elements denote the number of students in each process. Out of this initial stock, certain numbers, $(I-\hat{h})s$ say, where I denotes the unit matrix and $\hat{h}$ denotes a diagonal matrix of annual survival rates appropriate to the ages of students, die within the year. Out of those who live to the end of the year, a part $\hat{p}\hat{h}s$, say, continue in the system and the remainder, namely $(I-\hat{p})\hat{h}s$ are graduate leavers at the end of the year. Each element of p, the vector from which the diagonal matrix $\hat{p}$ is derived, denotes the probability that a surviving student who was in process j this year will be found in process $j+1$ next year. Evidently $p_j=1$ for compulsory processes and $0 \leqslant p_j \leqslant 1$ for the voluntary processes that succeed the years of compulsory education. Thus we can partition s as follows

$$s = (I-\hat{h})s+\hat{h}s$$
$$= (I-\hat{h})s+\hat{p}\hat{h}s+(I-)\hat{p}\hat{h}s \tag{1}$$

The second and third terms on the right-hand side of (1) can now be given a different notation. The second term, which represents the numbers of students who stay in the various processes, can be expressed in terms of next year's initial stock of students, Es say, where the operator E advances by one time unit the variable to which it is applied. Thus

$$\hat{p}\hat{h}s = JEs \tag{2}$$

where J denotes a matrix with ones in the diagonal immediately above the leading diagonal and zeros everywhere else. We cannot simplify (2) because both $\hat{p}$ and J are singular matrices; in the case of $\hat{p}$ because everyone in the last process is a graduate leaver from the system at the end of the year so that the final element of p is zero; and because J has non-zero elements only in a single diagonal which is not the leading one.

The third term on the right-hand side of (1) is simply the vector of graduate leavers at the end of the year. If we denote this vector by g, we can write

$$(I-\hat{p})\hat{h}s = g \tag{3}$$

Thus we can express graduate leavers in terms of the initial stock of students; but from (3) we cannot express s in terms of g: $(I-\hat{p})$ is a singular matrix because in the years of compulsory education the p_j are equal to one.

If we substitute from (2) and (3) into (1) and simplify, we obtain

$$s = \hat{h}^{-1}JEs + \hat{h}^{-1}g$$
$$= Ts + T\Delta s + \hat{h}^{-1}g \tag{4}$$

where $T \equiv \hat{h}^{-1}J$ and $\Delta \equiv (E-1)$ denotes the first-difference operator. Equation (4) is immediately recognized as the flow equation of an open, dynamic input–output model in which output levels are the elements of s and final products are the elements of $\hat{h}^{-1}g$ (not simply the elements of g because provision must be made for those who will die in the year). A peculiarity of this particular input–output model is that the capital co-efficient matrix (the premultiplier of Δs) is the same as the current coefficient matrix (the premultiplier of s).

Following the argument of Stone and Brown [157] we can rewrite (4) in the form

$$s = \sum_{\theta=0}^{\nu-1} T^\theta E^\theta \hat{h}^{-1}g \tag{5}$$

where ν denotes the number of processes and it is understood that $T^0 \equiv I$. The summation in this case only extends to $\nu-1$ and not to infinity: T is strictly upper triangular and so is equal to the null matrix when raised to any power $\theta > \nu-1$.

Throughout this paper I shall confine myself to simple cases of a more general problem so as to avoid complications which I believe can be handled fairly easily but would nevertheless blur the outlines of the model I am proposing with technical details. In the present instance it is clear that J is an unduly simple representation of the movement of students from process to process and that allowance must be made for the existence of more or less parallel streams and for variations in the time spent by different students in different processes. When this is done the general conclusions remain and it is more apparent than in my simple version that the matrix T is largely a reflection of learning times, which might be altered considerably by technical changes in the system.

Equation (5) shows the relationship between present numbers of students in the different processes and future numbers of graduate leavers from the different processes. Thus to any composition of these leavers over the coming $\nu-1$ years there corresponds a present composition of students. Provided that each future graduate vector can be expressed by an equation of the form of (3), in which p may, however, be a function of time, then the derived vector of present student numbers will equal the actual vector of present student numbers.

3. BIRTHS AND SURVIVALS

Let us now turn to the exogenous component in the supply of raw material to the educational system: the birth and death process that determines the evolution of the population. A peculiarity of the educational system, regarded as a set of productive processes, is that it is supplied exogenously with a stream of raw material in the form of young children. As, with time, this raw material becomes increasingly fabricated in the higher processes of the system, purely demographic factors, which can reasonably be regarded as exogenous in an educational model, give way, at least in some degree, to decisions made by parents, educators and the students themselves. These influences, which come into operation at the end of the period of compulsory education, are usually referred to as the demand for places and will form the subject of the next section.

In practice, it would be natural to take over detailed demographic projections compiled, independently of the educational model, in such a way that allowance was made for anticipated changes in specific birth and death rates and in rates of international migration. For theoretical purposes, on the other hand, it is convenient to have a simple representation of the birth and survival process in which many refinements are ignored. In this case the process can be represented, as shown in Leslie [97], by a square co-efficient matrix, H say, in which the age-specific birth rates appear in the first row and the age-specific survival rates appear diagonally immediately below the leading diagonal. If n denotes a vector in which the population is ordered by years of age, then we can write

$$E^\theta n = H^\theta n \tag{6}$$

for $\theta = 0, 1, 2, \ldots$.

For our purposes it is convenient to write (6) in partitioned form so as to isolate the population of student age. Let us denote by the suffixes 1, 2, 3, the age groups before, during and after education. If we express this year's population in terms of last year's population, we have

$$
\begin{bmatrix} n_1 \\ \cdots \\ n_2 \\ \cdots \\ n_3 \end{bmatrix} = \begin{bmatrix} H_{11} & \vdots & H_{12} & \vdots & H_{13} \\ \cdots & \vdots & \cdots & \vdots & \cdots \\ H_{21} & \vdots & H_{22} & \vdots & 0 \\ \cdots & \vdots & \cdots & \vdots & \cdots \\ 0 & \vdots & H_{32} & \vdots & H_{33} \end{bmatrix} \begin{bmatrix} E^{-1} n_1 \\ \cdots \\ E^{-1} n_2 \\ \cdots \\ E^{-1} n_3 \end{bmatrix} \tag{7}
$$

from which it follows that the population of student age, n_2, is given by

$$ n_2 = H_{21} E^{-1} n_1 + H_{22} E^{-1} n_2 \tag{8} $$

This equation simply states that the numbers in each student age-group at the beginning of this year are equal to last year's survivors from the preceding age-group.

A certain proportion of the population in each student age-group will be students. If we represent these proportions by the elements of a vector q, then

$$
\begin{aligned}
s &= \hat{q} n_2 \\
&= \hat{q}(H_{21} E^{-1} n_1 + H_{22} E^{-1} n_2)
\end{aligned} \tag{9}
$$

By applying the operator E to this expression we can generate student numbers for successive years in the future.

The elements of q are functions of the elements of p. For years of compulsory education, $p_r = q_r^{\scriptscriptstyle\square} = 1$. For the first stage of voluntary education, $q_s = p_s \leqslant 1$. For the second stage of voluntary education, $q_t = p_t p_s \leqslant p_s$, and so on. If the elements of p change through time, then the elements of q depend on elements of past p's. For example, $q_s = E^{-1} p_s$ and q_{s+1}, appropriate to the second year of the first voluntary process, is equal to $E^{-2} p_s$ on the assumption that all who enter the process stay to the end, that is $p_{s+1} = 1$. Again if the first voluntary process lasts two years then, on the same assumption, $q_t = E^{-1} p_t E^{-3} p_s$ and $q_{t+1} = E^{-2} p_t E^{-4} p_s$, and so on.

Thus, given any evolution of the elements of p, and therefore q, and the evolution of the population given by H and an initial population vector, $\bar{n}$ say, we can generate future values of s. If the future g's in (5) have been generated from the same information, the values of $E^\theta s$, $\theta = 0, 1, 2, \ldots$, derivable from (5) and from (9) will, of course, be the same.

In planning future education it cannot be assumed that p remains constant. Let us therefore consider the question of how it might change.

4. THE DEMAND FOR PLACES

Even if we restrict ourselves to the years of compulsory education, the demand for places may show itself in a preference for certain types of school:

THE DEMAND FOR PLACES 111

private schools or religious schools, say, rather than public (state) schools. Here we shall ignore this aspect of the problem and concentrate on the demand for places in the usual sense of the term: a demand for places in institutions of further education at the end of the years of compulsory education. This kind of demand may start with a decision to stay on to sixth form at school, then to go to a college or university and so on.

While, for short-run purposes, estimates of the future demand for places made by trend extrapolation may give acceptable results, it is desirable, as in other types of projection, to examine the forces which determine future demands. The suggestion made here is that higher education should be regarded as a series of epidemic processes [13] in which changes in the demand for places depend, in part, on the number infected and so liable to infect others and, in part, on the number not yet infected and so available to catch the infection.

Let us continue, as before, to make the assumption that there is a single sequence of processes of higher education, the first of which can be entered at the end of compulsory education, the second on passing successfully out of the first and so on. An example involving three processes is: staying on to sixth form, going on to undergraduate studies at a university, staying on there for postgraduate studies. Let us denote by p_1 the proportion of this year's school-leavers who stay on to sixth form next year, by p_2 the proportion of sixth-form leavers who go on to undergraduate studies, by p_3 the proportion of university graduates who continue for postgraduate studies, and let $p^* \equiv \{p_1 p_2 p_3\}$. Then we might describe the evolution of student numbers in these three voluntary processes by the following set of difference equations

$$\Delta p_1 = a_1 p_1 (1 - b_1 - p_1) \tag{10}$$

$$\Delta p_2 = a_2 p_2 (1 - b_2 - p_2) \tag{11}$$

$$\Delta p_3 = a_3 p_3 (1 - b_3 - p_3) \tag{12}$$

or, in general, by the matrix equation

$$\Delta p^* = \hat{a} \hat{p}^* \, (i - b - p^*) \tag{13}$$

Equation (10) states that Δp_1, the change from one year to the next in the proportion of students going on to voluntary process 1, is related by a factor of proportionality a_1 to the product of the proportion p_1 who attended this process last year and the proportion $(1 - b_1 - p_1)$ who did not attend it last year but had the ability and, in happier circumstances, would have had the wish to do so. If eventually all were to have the ability and the wish to do so, then we should have $b_1 = 0$. In a similar way, (11) states that Δp_2, the change from one year to the next in the proportion of students going on from voluntary process 1 to voluntary process 2, is related by a factor of proportionality, a_2, to the product of p_2 and $(1-b_2-p_2)$. And so on. From (13) we can immediately write

$$Ep^* = p^*[i+a(i-b-p^*)]$$ (14)

which expresses next year's value of p^* in terms of this year's value of p^* and the parameters a and b.

This system of equations gives rise to growth curves which carry the proportions in each process from initial positive values, the elements of a vector $\bar{p}^*$ say, to final values, the element at stage r being equal to $\prod_{j=1}^{r}(1-b_j)$. If we start from sufficiently small positive values of all the p_j, these growth curves will all be sigmoid in shape. The general evolution of the system is perhaps easier to appreciate if the equations are written in differential form, that is if Δp_j is replaced by $\dot{p}_j \equiv dp_j/dt$. With this change (13) represents a set of linear differential equations in the reciprocals of the p_j. Each member of this set is the differential equation of a logistic process whose solution is

$$p_j = \frac{1-b_j}{1+\left(\dfrac{1-b_j}{\bar{p}_j}-1\right)e^{-a_j(1-b_j)t}}$$ (15)

where $\bar{p}_j$ is the initial value of p_j at time $t=0$.

We thus see that the elements of q in (9), being products of the elements of p, turn out, on the present assumptions, to change through time as the product of one or more logistic processes. The problem of dating is easily solved by adjusting the value of t in the different terms of the logistic product. A corresponding result is obtained by the use of finite differences as in (13). In the fullness of time the contagions will spread until everyone susceptible to them is infected.

The main interest of this scheme lies in the fact that the Δp_j are not constant or even monotonic but vary from low values when p_j is small through intermediate higher values to low values again when p_j is large. If, therefore, the real world is even approximately like this scheme, we should recognize that past trends may be a poor guide to future changes in the demand for places. The changes may suddenly start to grow much faster only to die down again as the greater part of the demand is provided for.

There is one point to bear in mind in applying this scheme. The available evidence [178] shows that different social classes are at very different stages of educational penetration. For example, of children born in Britain in 1940–41 to fathers with different occupations, the proportion attending full-time higher education fell steadily from 45 per cent among higher professionals to 2 per cent among the semi- and unskilled. This suggests that it would be desirable to treat different social classes separately as far as possible.

5. THE INFLUENCE OF ECONOMIC PROSPECTS

In the preceding section we examined a possible method of simulating the demand for places at the various stages of higher education, a demand which, as the Robbins Committee has emphasized, ought, as far as possible, to determine the broad pattern of educational facilities available. In that section no attempt was made to model the various processes in detail, to consider, let us say, not just the demand for a place in a university but the demand for a place in a university school of physics, biology, engineering and so on. In trying to repair this defect we must replace the single series of stages with a series of branching processes in which from a given stage it is possible to choose one of many processes at the next stage.

Let us concentrate on a single example of this branching process: the choice of alternative subjects at the university. Here the epidemic model, which was used to determine the total number of university entrants, is not appropriate; what is needed is to give weight to the relative attractiveness of different courses as judged by their difficulty and prospects rather than to the contagious effect of the example of others. One way to do this is to make the choices in the next period depend partly on the choices of this period and partly on a ranking of the different courses that expresses the outcome of an objective assessment of prospects. Thus, if we denote by r_j a vector whose elements are the present proportions of students at stage j who attend the different courses and by r_j^* a vector whose elements are the corresponding proportions as they would be on an objective assessment of prospects, we could write

$$Er_j = \alpha_j r_j + (1-\alpha_j) r_j^* \tag{16}$$

where α_j and the elements of r_j and r_j^* are non-negative and not greater than one and where $i'r_j = i'r_j^* = 1$. This relationship states simply that future r_j will move from r_j to r_j^*. The time taken to adapt depends on α_j, a low value implying a high rate of adaptation. Equation (16) will be recognized as one form of the basic equation used in models of the learning process [29].

The vector r_j^* may be expected to change through time with changing assessments of the different courses so that students are continually adjusting themselves to a changing terminal state. It seems reasonable to suppose that an important factor making for changes in r_j^* is the changing economic prospects of the various branches of study, since these prospects, whether rightly or wrongly assessed, are likely to loom large in the minds of parents, educators and the students themselves when the choice of a course is being made. Though no one in his senses would claim that education exists simply to enable individuals to acquire economically useful skills, at the same time it should be recognized that most people have to earn their living and it is desirable that their education should prepare them to do this to their satisfaction. This means that we need what light we can get on the future prospects of different occupations and the way in which satisfactory performance in these occupations is related to educational achievement.

This is one of the points at which a model of the educational system links up with a general model of the economy that seeks to project the output levels of different branches of economic activity, the techniques that will be employed in them and the consequent size and composition of the human establishment they will try to maintain.

6. ECONOMIC INPUTS INTO EDUCATION

The purpose of calculating present and future values of s is to enable us to take appropriate action in a particular long-term investment project: investment in education. If we want to realize in the future the graduate mixes determined by the evolution of the demand for places and the needs of the economic system, we must make sure that intermediate activities are, in the meantime, carried out on a sufficient scale. To see what is implied by this we must calculate the primary and intermediate economic inputs, teachers, buildings, equipment and supplies, into the various processes of the system. These depend in the first place on activity levels and in the second place on educational technology. In this section we shall consider the first of these influences.

Let X denote a matrix whose rows relate to different economic inputs and whose columns relate to different educational processes. The units in which these inputs are measured must be uniform for each input but will normally vary from input to input: numbers of teachers, square feet of buildings, gallons of ink, etc. The elements in a column of X express the economic input structure of a particular process. If we divide the elements in each column by the corresponding activity level we obtain a coefficient matrix. Denoting this matrix by U, we have

$$U = X\hat{s}^{-1} \tag{17}$$

We now see that educational technology is summarized not in one matrix, T, but in two matrices, T and U. Changes in T mainly reflect changes in the normal time assigned to different learning processes; changes in U reflect changes in the economic inputs into these processes. The two changes will often be linked: a new curriculum may reduce learning time but also require different equipment or a new type of teacher.

If we denote the row sums of X by x, then from (17) it follows that

$$x = Us \tag{18}$$

If the elements of U are constant then we can calculate the value of x corresponding to any given value of s. From (5) or (9) we can calculate the required value of s and in a similar way we can calculate the required value of $E^\theta s$, $\theta = 1, 2, \ldots$. We can thus calculate the corresponding $E^\theta x$ and compare these with the actual value of x, say x^*, and existing intentions for $E^\theta x$. If the x^* series is less than the x series, the plan cannot be realized as it stands and it will be necessary either to postpone it for the minimum time needed to build up the required economic inputs or to adopt some stop-gap measures, usually involving a temporary lowering of standards, which will allow it to go forward with less than its normal complement of economic inputs.

7. TECHNOLOGICAL CHANGE IN EDUCATION

The purpose of this section is not to discuss possible changes in educational technology but simply to show that any assumptions about future possibilities can be introduced into the model and so affect the outcome of the calculations. This result follows immediately from the argument in [157]. If T does not remain constant, we may denote the value of this matrix θ years from now by $E^\theta T$. In this case (5) is replaced by

$$ s = \left[I + \sum_{\theta=1}^{\nu-1} \left(\prod_{\tau=1}^{\theta} E^{\tau-1} T \right) E^\theta \right] g \qquad (19) $$

If at the same time U does not remain constant and we denote the value of this matrix θ years from now by $E^\theta U$, then evidently the value of x in (18) is unchanged but the value of $E^\theta x$ for $\theta > 0$ is changed.

8. THE SUPPLY OF ECONOMIC INPUTS

It is convenient to discuss the supply of economic inputs in the following order: (a) intermediate inputs (supplies), that is ink, paper, books, etc.; (b) capital goods (buildings and equipment); and (c) labour (mainly teachers).

As regards (a), little need be said here. Requirements for supplies are related to activity levels, the elements of s, by coefficients, elements of U. It is hard to imagine circumstances in which these inputs could pose a serious problem.

As regards (b), requirements for buildings and equipment depend on activity levels and so investment in buildings and equipment depends on changes in activity levels except in so far as existing buildings and equipment have been allowed to get into a substandard condition. Investment in capital goods in education is similar to investment in capital goods in any other economic activity, a model for which is described in [30].

As regards (c), an obvious special circumstance exists: the educational system itself produces the supply of potential teachers and must ensure that a sufficient number of these are attracted to take up teaching as a profession. The position can be examined by writing down equations for the demand for and supply of teachers.

9

The demand for teachers depends on the activity levels of the different educational processes. Thus we can write

$$Ee_d = AEs \tag{20}$$

where the elements of e_d relate to the demand for different kinds of teacher and A denotes a partition of U, namely the set of rows relating to teachers in the various educational processes.

The supply of teachers depends on two factors: (i) the number in the existing stock of teachers who remain active after allowance has been made for wastage due to change of job, retirement or death; and (ii) the number of graduate leavers who take up teaching as a career. Thus we can write

$$Ee_s = (I - \hat{c})e_s + Bg \tag{21}$$

where the elements of e_s relate to the supply of different kinds of teacher, the elements of $\hat{c}$ relate to the wastage rates of different kinds of teacher and B denotes a matrix whose elements show the proportion of different kinds of graduate leaver who take up different kinds of teaching.

If everything is to go smoothly we must be able to equate demand and supply. If demand tends to exceed supply, then we must either: (i) reduce teacher–student ratios, that is reduce the elements of A; (ii) reduce wastage rates, that is reduce the elements of c; or (iii) increase the proportion of graduate leavers who return to teaching, that is increase the elements of B. An improvement in the status and pay of teachers can be expected to contribute to all these changes and differential improvements as between different classes of teacher can be expected to reduce specific imbalances.

9. THE NEXT STEPS

In trying to apply this model, the next step is to compile the data necessary to estimate its various parameters at the same time modifying and improving the constituent relationships. When this has been done we shall have a computable model designed to show the consequences of demographic changes and of giving the demand for places its head.

When this stage is reached we are likely to find that we shall have to consider entirely new educational processes and decide on the processes that are actually to be used in the future. This is one reason why eventually we shall have to adopt a programming formulation of the model. But the intermediate input–output stage is not to be despised since it gives us a manageable, if ambitious, statistical programme and provides the empirical groundwork for the further development of the model.

Finally, the outcome of the model must be related to the state of the economy as a whole. I have described the model as demand-determined but the economics of this demand are extremely shadowy since I have tacitly assumed that, as at present, the families of most students will bear directly only a small part of the costs of education. On the whole this seems to me a highly desirable state of affairs but it only throws back on to those responsible for general social and economic policy the decision of how much of the national resources should be devoted to education. This brings us to an area of social cost-benefit analysis in which market values are a necessary but by no means sufficient ingredient.

X

SOCIAL ACCOUNTS AT THE REGIONAL LEVEL: A SURVEY

1. INTRODUCTION

The purpose of this paper is to set out in an orderly manner methods of regional description and analysis which centre round the social accounts. We may take the national accounts as a standard of reference and apply this form of accounting to a set of interconnected regions. If we consolidate a set of accounts until there are only two accounts left, we obtain a pair of identical transactions and so a single indicator of activity. As we increase the number of accounts, the number of transactions distinguished increases. At an early stage in the elaboration of the accounting system the national accounts will be reached. In these, the transactions of all transactors in the economy are classified according as they relate to one of the three fundamental forms of economic activity: production, consumption and accumulation. In an open economy it will be necessary to add a fourth account containing transactions with the rest of the world. At this stage we may again increase the number of accounts; and this may be done in various ways. A particularly fruitful way is to subdivide the account relating to production so as to account separately for the different branches of productive activity which make up production as a whole. In this way a set of input–output accounts is introduced into the national accounting framework. Having reached this stage we could increase the number of accounts still further and so obtain a more detailed system of social accounting.

If these ideas are applied to a set of regions, a certain amount of information will be provided at each stage in the progression, the amount increasing as the progression continues. At each stage, the information available will permit certain questions to be answered and there will be certain methods which are most efficient for the purpose. In this paper an attempt is made to keep data and methods in step and to relate both to the kind of hypothesis in which the investigator is interested.

The order of the paper is as follows. In section 2 the case is considered in which there is a single indicator for each region. This is an important case since it frequently happens that the only accounting information available relates to total production or income payments in the states, provinces, counties or other regions of a country. It is suggested that the methods of analysis in this case can conveniently be brought together within the framework of the analysis of variance and covariance.

In section 3 we take a step forward in the elaboration of the accounting system and consider a set of regional accounts drawn up like national accounts. A definite form is given to this system and various rearrangements to which it can be subjected are described. An accounting system implies certain relationships among the constituent transactions, and in section 4 the use of these relationships is considered:

(i) for the purpose of indirect estimation; and
(ii) for the adjustment of a complete set of direct estimates to provide a fully balanced set of accounts.

In the national accounts there is an account for the rest of the world. In the corresponding regional accounting system it is convenient to introduce a set of accounts for all regions and the rest of the world. In section 5 it is shown that these accounts can be used to accommodate sectors of the economy, such as the central government, which have no well defined location.

In section 6 the foregoing ideas are illustrated by a set of regional accounts relating to the United Kingdom in 1948. In these accounts each of the twelve regions is characterized by a set of fourteen transactions.

Accounts for regions, like accounts for different time periods, raise the question of adjustment to a common price structure. Conventional methods are based on pairwise comparisons but these break down if more than two countries are involved. Means of producing a consistent set of comparisons are considered in section 7.

At this point we turn from description to analysis based on a set of regional accounts. Thus in section 8, the question of classifying regions is considered when no clear views about criteria for classification are available. For this purpose the regions are considered as points in the space of the transactions and the distance between them is measured. An attempt is then made to assign them to clusters in such a way that the average distance within clusters is small compared with the average distance between clusters. This method has much in common with the analysis of a set of variables into orthogonal components or factors.

Section 9 is concerned with the use of regional accounting data for the purpose of investigating individual relationships. As an example an attempt is made to estimate the parameters of a saving function based on the accounts of section 6.

From partial analysis we pass to complete models. In section 10 the possibility of establishing a regional model on national accounting lines is examined. This model is formally identical to the open static input–output model except that in this case it is necessary to replace the assumption of proportionality by the more general assumption of linearity.

In section 11 another step is taken in the elaboration of the accounting system by subdividing the production account for each region on an industry basis. This leads to regional input–output within a national accounts framework. It gives rise to two different kinds of regional input–output model. The first, a centrally connected regional model, is described in section 12. The second, a model with variable cost structures and bilaterally fixed trading patterns, is described in section 13. Section 14 is devoted to local and regional input–output studies in various countries.

The methods so far described enable many different analyses to be made. While these are likely to be useful in development planning, they do not reduce the planning problem to model form. For this purpose it is necessary to step into the field of programming. No more is done here than to indicate the existence of a great deal of recent work concerned with the locative aspects of programming.

The paper ends with a brief appendix on the estimates contained in the set of regional accounts described in section 6.

2. SINGLE INDICATORS FOR EACH REGION

At the present time it is unusual to find the social accounts of a country subdivided on a regional basis; the most that is usually available for regions is a limited number of income or product totals. For example, [183] contains data on income payments by states for the United States of America and [37] contains similar data for the provinces of Canada. Such information has been extensively discussed and analysed, especially for the United States, in such publications as [68, 182]. Sometimes information is available for still smaller regions: an example is Lancaster's study of county income estimates for seven south-eastern states [90] which also contains references to a number of similar studies.

This information, and especially that for very small areas, is frequently used in the assessment of local markets. The first requirement, namely a measure of the size of these markets, is met by the estimates themselves. Subsequent requirements usually involve a comparison of different regions, to establish how far they are homogeneous, how far they are subject to different trend or cyclical influences and with what factors, such as changes in population or differences in industrial composition, these differential movements are associated. In order to carry out these analyses use must be made of the tools of mathematical statistics. Although it is necessary to use different tools for different purposes, it would seem that in the present case the appropriate tools can largely be brought together within the framework of the analysis of variance and covariance.

Suppose that we have measures of income payments in each of R regions for each of T years. If we denote the measure for region r in year t by y_{rt} then we can decompose the sum of squares $\sum_r \sum_t y_{rt}^2$ into three independent parts, one associated with the general mean of the observations, one associated with the deviations of the regional means from the general mean and one associated with the deviations of the individual observations for the regions about their respective regional means. In terms of regression theory that is equivalent to the model

$$y_{rt} = \alpha + \beta_r + \epsilon_{rt} \tag{1}$$

where α denotes the true mean of the RT observations, β_r denotes the excess over α of the true mean for region r and ϵ_{rt} denotes a true residual. Provided that the ϵ_{rt} are independently distributed with constant variance then α and the β_r can be estimated by minimizing the sum of squares, $\sum_r \sum_t \epsilon_{rt}^2$, subject to the condition that $\sum \beta_r = 0$. This problem is formally similar to the method of analysing seasonal variations proposed in [150]. A good treatment of estimation subject to constraints can be found in [3].

In practice the successive annual measures for a region cannot be regarded as random drawings from a fixed probability distribution since trend and cyclical influences are usually at work to ensure certain systematic tendencies in every sequence of observations. Some allowance can be made for these influences by introducing a factor which varies with the years in addition to the factor which varies with the regions. Thus (1) becomes

$$y_{rt} = \alpha + \beta_r + \gamma_t + \epsilon_{rt} \tag{2}$$

where the γ_t relate to the deviations of the means for individual years about the general mean. In this case the sum of squares, $\sum_r \sum_t \epsilon_{rt}^2$, must be minimized subject to the two conditions $\sum_r \beta_r = \sum_t \gamma_t = 0$. The corresponding analysis of variance takes the usual form for a two-way classification.

This type of analysis may be extended by introducing real variables as well as the dummy ones associated with regions and years. Perhaps the simplest case arises when we consider the possibility that regions are distinguished partly by differential trends. In this case the basic equation becomes

$$y_{rt} = \alpha + \beta_r + \gamma_t + \delta_r t + \epsilon_{rt} \tag{3}$$

where the term $\delta_r t$ measures the trend in the observations for region r. Further variables may be introduced in this way but it is then usually desirable to measure them in such a way that they are independent of the dummy variables. For example climatic conditions may be important in accounting for the observed differences in the y_{rt} but it is desirable that they should be measured as deviations from regional norms since otherwise there will be a confusion between their influence and that of the unspecified factors that contribute to the differences in regional means.

3. REGIONAL ACCOUNTS

A fuller analysis of regional differences and inter-relationships requires that the single indicator of regional activity be decomposed and extended so as to show the transactions within and between regions in some detail. A conceptual framework for this type of information is provided by a system of regional accounts. These in turn can be built up from two building blocks, one relating to intra-regional and the other to inter-regional transactions.

Consider n regions, $j, k = 1, \ldots, n$, which share a common currency so that in a first approximation there is no adjustment to be made to the unit in which transactions are measured. Suppose that for each region a set of accounts, like national accounts, is established, one relating to production, one relating to consumption and one relating to accumulation. If the n regions form a closed economic world, there is no need to provide each with an external account since every external transaction of a region flows between one of the accounts of that region and one of the accounts of one of the other regions.

If we set out the accounts in the familiar matrix form with each row and column pair representing an account, and if we place the outgoings in columns and the incomings in rows, then the intra-regional building block for region j might take the form:

TABLE 1   THE INTRA-REGIONAL BUILDING BLOCK

	P	C	K
P	0	C_{jj}	V_{jj}
C	Y_{jj}	0	0
K	D_{jj}	S_{jj}	0

The entries in the first row of this table are the incomings into the production account of region j and consist of consumption, C_{jj}, and investment, V_{jj}. The entries in the first column are the outgoings of this account and consist of income payments to domestic factors of production, Y_{jj}, and depreciation, D_{jj}. Correspondingly, the entry in the second row, Y_{jj}, is the income available for consumption and the entries in the second column are consumption, C_{jj}, and saving, S_{jj}. Finally, the entries in the third row and column represent the incomings, D_{jj} and S_{jj}, and the outgoings, V_{jj}, of the capital transactions account of region j.

If more than one region is considered a table of the above form will be incomplete since it relates only to intra-regional transactions. The complementary building block for inter-regional transactions might take the form:

TABLE 2   THE INTER-REGIONAL BUILDING BLOCK

X_{jk}	0	0
Y_{jk}	G_{jk}	0
0	0	B_{jk}

In this table the rows relate, as before, to the three accounts of region j while the columns relate to the three accounts of region k. Thus X_{jk} denotes the exports from j to k, Y_{jk} denotes the factor income payments received by j from k, G_{jk} denotes the gifts and grants received by j from k and B_{jk} denotes the amount which j borrows from k.

If n regions form a closed economy, we can construct a complete accounting system with n building blocks of the first type and $n(n-1)$ of the second type. Thus, if $n=3$, the transactions within and between the regions can be arranged as in table 3.

TABLE 3   ACCOUNTS FOR THREE REGIONS ORDERED BY REGION AND TYPE OF ACCOUNT

	Region 1			Region 2			Region 3		
Region 1	0	C_{11}	V_{11}	X_{12}	0	0	X_{13}	0	0
	Y_{11}	0	0	Y_{12}	G_{12}	0	Y_{13}	G_{13}	0
	D_{11}	S_{11}	0	0	0	B_{12}	0	0	B_{13}
Region 2	X_{21}	0	0	0	C_{22}	V_{22}	X_{23}	0	0
	Y_{21}	G_{21}	0	Y_{22}	0	0	Y_{23}	G_{23}	0
	0	0	B_{21}	D_{22}	S_{22}	0	0	0	B_{23}
Region 3	X_{31}	0	0	X_{32}	0	0	0	C_{33}	V_{33}
	Y_{31}	G_{31}	0	Y_{32}	G_{32}	0	Y_{33}	0	0
	0	0	B_{31}	0	0	B_{32}	D_{33}	S_{33}	0

In this table the intra-regional transactions appear in the blocks placed diagonally from the top left-hand to the bottom right-hand corner, and the inter-regional transactions appear in the remaining blocks.

Before we turn to some of the special problems which arise in regional accounting, it is worth noting that we can rearrange the order of the accounts in the preceding table so as to concentrate each type of transaction in a single block. This is done by ordering the rows and columns in the first place by type of account and in the second place by region and not in the reverse order as was done above. In the table below the first three rows and columns relate to production accounts, one in each region, the second three to income and outlay accounts and the last three to capital transactions accounts.

TABLE 4   ACCOUNTS FOR THREE REGIONS ORDERED BY TYPE OF ACCOUNT
AND REGION

		P			C			K		
P	1	0	X_{12}	X_{13}	C_{11}	0	0	V_{11}	0	0
	2	X_{21}	0	X_{23}	0	C_{22}	0	0	V_{22}	0
	3	X_{31}	X_{32}	0	0	0	C_{33}	0	0	V_{33}
C	1	Y_{11}	Y_{12}	Y_{13}	0	G_{12}	G_{13}	0	0	0
	2	Y_{21}	Y_{22}	Y_{23}	G_{21}	0	G_{23}	0	0	0
	3	Y_{31}	Y_{32}	Y_{33}	G_{31}	G_{32}	0	0	0	0
K	1	D_{11}	0	0	S_{11}	0	0	0	B_{12}	B_{13}
	2	0	D_{22}	0	0	S_{22}	0	B_{21}	0	B_{23}
	3	0	0	D_{33}	0	0	S_{33}	B_{31}	B_{32}	0

The accounting system set out in tables 3 and 4 presupposes an unusual amount of information about transactions between regions. Thus if the regions are countries it is likely that the X_{jk}, that is the imports and exports between pairs of countries, will be known. But if the regions are geographical subdivisions of a country it is most unlikely that there will be information about the X_{jk}. If we move to the Y's or the G's or the B's it is even unlikely that such information will be available for countries. Accordingly it is important to be able to arrange the accounts so that this detailed information is unnecessary. This can be done by introducing a set of accounts for all regions together and ensuring that all inter-regional transactions flow between an account in one of the regions and one of the new accounts for all regions. In this way, table 3 would be changed as in table 5.

This table is obtained from table 3 by the following operations. First, table 3 is bordered by three rows and three columns in respect of the three accounts relating to all regions. Second, the intra-regional transactions remain unchanged in the diagonal blocks. Finally, the inter-regional transactions in the off-diagonal blocks of table 3 are removed from those blocks, added together row-wise and column-wise, and entered in one of the new blocks which connect a single region with all the others taken together. These new blocks are similar in form to the off-diagonal blocks of table 3, the only difference being that the elements in the new blocks are the sums of elements in the old ones. For example $X_{12} + X_{13}$ denotes the total exports from region 1 to the other two, so that with this arrangement no pairwise inter-regional information is needed.

It will be noticed that in table 5 each of the nine accounts relating to individual regions balances exactly as in table 3. But the first two accounts for all regions only balance when the entry $-\Sigma \Sigma Y_{jk}$ is made. This entry is $k\ j\neq k$ simply the negative of total inter-regional payments of factor income.

If the ordering of the accounts in table 5 is changed so as to concentrate each type of transaction into a single block the result is as in table 6.

In this table the first three row and column pairs represent the production accounts of the three regions. The fourth row and column pair contains in its row, the imports of the three regions and, in its column, the exports of the three regions. The remaining rows and columns are arranged on a similar principle.

Tables 5 and 6, though less demanding of inter-regional information than tables 3 and 4, can be simplified still further. This is done by applying a process of 'netting' to all inter-regional flows. In each region, outgoing flows to other regions may be netted off the corresponding incoming flows. To represent this, two additions are needed to the notation used so far. For region j, the net flow from other regions is denoted by a single subscript. Thus $X_{12} + X_{13} - X_{21} - X_{31} \equiv X_1$. When this operation is performed on the Y's to form the Y_j, it is clear that the production and the income and outlay accounts will be thrown out of balance unless the factor income payments to other regions are added to the diagonal elements Y_{jj}. Since ΣY_{kj} is equal to domestic income, or net domestic product at factor cost, k of region j, this sum is denoted by Y_{dj}. Thus for example $Y_{11} + Y_{21} + Y_{31} \equiv Y_{d1}$.

In its net form, table 6 appears as in table 7.

In this highly simplified set of accounts, the following relationships emerge. In the regional production accounts, the gross domestic product, $Y_{dj} + D_{jj}$, is matched by net exports, consumption and gross investment in the region, $X_j + C_{jj} + V_{jj}$. In the regional income and outlay accounts, consumption and saving in the region, $C_{jj} + S_{jj}$, are matched by income to the factors of production in the region and net gifts received, $Y_{dj} + Y_j + G_j$. Finally in the regional capital transactions accounts, gross investment, V_{jj}, is matched by depreciation, saving and net borrowing, $D_{jj} + S_{jj} + B_j$.

In table 7 each account balances and in addition it is known, for example, that $\Sigma X_j \equiv 0$. Accordingly there are twelve independent relationships j connecting the twenty-seven flows.

TABLE 5   ACCOUNTS FOR THREE REGIONS ORDERED BY REGION AND TYPE OF ACCOUNT WITH NO PAIRWISE TRANSACTIONS

0	C_{11}	V_{11}	0	0	0	0	0	0	$X_{12}+X_{13}$	0	0
Y_{11}	0	0	0	0	0	0	0	0	$Y_{12}+Y_{13}$	$G_{12}+G_{13}$	0
D_{11}	S_{11}	0	0	0	0	0	0	0	0	0	$B_{12}+B_{13}$
0	0	0	0	C_{22}	V_{22}	0	0	0	$X_{21}+X_{23}$	0	0
0	0	0	Y_{22}	0	0	0	0	0	$Y_{21}+Y_{23}$	$G_{21}+G_{23}$	0
0	0	0	D_{22}	S_{22}	0	0	0	0	0	0	$B_{21}+B_{23}$
0	0	0	0	0	0	0	C_{33}	V_{33}	$X_{31}+X_{32}$	0	0
0	0	0	0	0	0	Y_{33}	0	0	$Y_{31}+Y_{32}$	$G_{31}+G_{32}$	0
0	0	0	0	0	0	D_{33}	S_{33}	0	0	0	$B_{31}+B_{32}$
$X_{21}+X_{31}$	0	0	$X_{12}+X_{32}$	0	0	$X_{13}+X_{23}$	0	0	0	0	0
$Y_{21}+Y_{31}$	$G_{21}+G_{31}$	0	$Y_{12}+Y_{32}$	$G_{12}+G_{32}$	0	$Y_{13}+Y_{23}$	$G_{13}+G_{23}$	0	$-\sum\limits_{k}\sum\limits_{j\neq k} Y_{jk}$	0	0
0	0	$B_{21}+B_{31}$	0	0	$B_{12}+B_{32}$	0	0	$B_{13}+B_{23}$	0	0	0

TABLE 6 ACCOUNTS FOR THREE REGIONS ORDERED BY TYPE OF ACCOUNT AND REGION WITH NO PAIRWISE TRANSACTIONS

0	0	0	$X_{12}+X_{13}$	C_{11}	0	0	0	V_{11}	0	0	0
0	0	0	$X_{21}+X_{23}$	0	C_{22}	0	0	0	V_{22}	0	0
0	0	0	$X_{31}+X_{32}$	0	0	C_{33}	0	0	0	V_{33}	0
$X_{21}+X_{31}$	$X_{12}+X_{32}$	$X_{13}+X_{23}$	0	0	0	0	0	0	0	0	0
Y_{11}	0	0	$Y_{12}+Y_{13}$	0	0	0	$G_{12}+G_{13}$	0	0	0	0
0	Y_{22}	0	$Y_{21}+Y_{23}$	0	0	0	$G_{21}+G_{23}$	0	0	0	0
0	0	Y_{33}	$Y_{31}+Y_{32}$	0	0	0	$G_{31}+G_{32}$	0	0	0	0
$Y_{21}+Y_{31}$	$Y_{12}+Y_{32}$	$Y_{13}+Y_{23}$	$-\sum_{k}\sum_{j\neq k}Y_{jk}$	$G_{21}+G_{31}$	$G_{12}+G_{32}$	$G_{13}+G_{23}$	0	0	0	0	0
D_{11}	0	0	0	S_{11}	0	0	0	0	0	0	0
0	D_{22}	0	0	0	S_{22}	0	0	0	0	0	0
0	0	D_{33}	0	0	0	S_{33}	0	0	0	0	0
0	0	0	0	0	0	0	0	0	0	0	0
0	0	0	0	0	0	0	0	0	0	0	$B_{12}+B_{13}$
0	0	0	0	0	0	0	0	0	0	0	$B_{21}+B_{23}$
0	0	0	0	0	0	0	0	0	0	0	$B_{31}+B_{32}$
0	0	0	0	0	0	0	0	$B_{21}+B_{31}$	$B_{12}+B_{32}$	$B_{13}+B_{23}$	0

TABLE 7   ACCOUNTS FOR THREE REGIONS IN NET FORM WITH NO PAIRWISE
TRANSACTIONS

0	0	0	X_1	C_{11}	0	0	0	V_{11}	0	0	0
0	0	0	X_2	0	C_{22}	0	0	0	V_{22}	0	0
0	0	0	X_3	0	0	C_{33}	0	0	0	V_{33}	0
0	0	0	0	0	0	0	0	0	0	0	0
Y_{d1}	0	0	Y_1	0	0	0	G_1	0	0	0	0
0	Y_{d2}	0	Y_2	0	0	0	G_2	0	0	0	0
0	0	Y_{d3}	Y_3	0	0	0	G_3	0	0	0	0
0	0	0	0	0	0	0	0	0	0	0	0
D_{11}	0	0	0	S_{11}	0	0	0	0	0	0	B_1
0	D_{22}	0	0	0	S_{22}	0	0	0	0	0	B_2
0	0	D_{33}	0	0	0	S_{33}	0	0	0	0	B_3
0	0	0	0	0	0	0	0	0	0	0	0

4. DIRECT AND INDIRECT ESTIMATION AND THE ADJUSTMENT
OF OBSERVATIONS

Tables 3 through 7 illustrate simple national accounting schemes applied to regions. They contain in each case a certain number of non-zero entries connected by a certain number of independent relationships. For a closed world of two or more regions, the numbers are as in table 8.

The number of flows is obtained by observing how many times flows of different kinds occur in each of the tables; for tables 5 and 6 the flow $\sum_{k} \sum_{j \neq k} Y_{jk}$ is not counted, and nor is its definition when the number of independent relationships is given. Only in tables 3 and 4 does the number of flows involve the square of n. Accordingly in adopting a form of presentation like tables 5, 6 or 7 in place of one like tables 3 or 4 we both lose a great deal of information and at the same time greatly simplify our requirements for data. The number of independent relationships is obtained from the consideration that entries in a closed system of n^* balancing accounts are connected by n^*-1 independent accounting relationships. In tables 3 and 4 there are $3n$ accounts and so $3n-1$ independent relationships. In tables 5 and 6 there are $3(n+1)$ balancing accounts and so $3n+2$ independent relationships omitting the definition of $\sum_{k} \sum_{j \neq k} Y_{jk}$. In table 7 all accounts balance and in addition it is known that $\sum_{j} X_j \equiv 0$. Since there are $3(n+1)$

TABLE 8   FLOWS AND INDEPENDENT RELATIONSHIPS FOR A
CLOSED WORLD OF $n > 1$ REGIONS

Tables	Flows	Independent relationships
Tables 3 and 4	$n(4n+1)$	$3n-1$
Tables 5 and 6	$13n$	$3n+2$
Table 7	$9n$	$3(n+1)$

accounts there are, as we have seen, an equal number of independent relationships.

In constructing accounting tables these relations can be used in either of two ways. The first, and simplest, is to make indirect or residual estimates of some of the flows which cannot be estimated directly. The second is to use the relationships to adjust an inconsistent set of direct estimates.

If there are m independent relationships connecting a set of variables then at most m of these variables can be estimated by combining the relationships with direct estimates of the remaining variables. There is a simple way of discovering whether any particular subset of m or fewer variables can be estimated in this way. If the accounts are represented by nodes and the entries are represented by branches, it is only necessary to connect the accounts by the m or fewer branches and observe whether the resulting diagram, or linear graph, contains any closed circuits, or loops. If it does not, the unknown variables can all be estimated residually; if it does, only those which remain after any loop has been broken by the removal of a branch can be estimated residually. If there are exactly m unknowns, the graph which corresponds to the possibility of complete residual estimation is simply connected, that is all nodes are connected and there are no loops. Such a graph is called a tree.

The adjustment of an inconsistent set of direct estimates is more difficult, but it is sufficiently important to be worth a mention here. Suppose that we had succeeded in making direct estimates of all the entries in table 3 but that they did not balance out. If we had done our best with the available data we might be willing to agree that we believed our direct estimates to be unbiased but that we should nevertheless revise them up or down by appropriate amounts in order to balance the table. If we thought that a particular direct estimate was substantially accurate we should not be willing to change it much and we might even be unwilling to change it at all. On the other hand if we had little confidence in a particular direct estimate we should be willing to revise it substantially. In some cases we should probably find that the direct estimates were linked; for example if an accurate total were divided into two parts we should not be willing to increase one part without at the same time reducing the other part correspondingly.

The following formulation is due to Durbin [51] and is a generalization of the one set out in [161]. Consider a vector x^* of type $m \times 1$ which contains unbiased estimates of another vector x. Suppose that the elements of x (the true values) are subject to μ linear constraints, that is

$$Gx = h \tag{4}$$

where G is a matrix of type $\mu \times m$ and rank μ, and h is a vector of type $\mu \times 1$. Let V^*, of order m and rank greater than μ, denote the variance matrix of the elements of x^*, and assume that any constraints satisfied by x^* are linearly independent of (4). Then the best linear unbiased estimator, x^{**}, of x is given by

$$x^{**} = x^* - V^* G'(GV^* G')^{-1}(Gx^* - h) \tag{5}$$

from which it follows that the elements of V^* need only be approximated up to a constant multiplier since any such constant will disappear in the matrix product $V^* G'(GV^* G')^{-1}$. The variance matrix of x^{**}, say V^{**}, is

$$V^{**} = V^* - V^* G'(GV^* G')^{-1} GV^* \tag{6}$$

It is the elements of the matrix V^* which have to be estimated subjectively but, provided that a positive amount of information can be supplied by those responsible for the direct estimates, it is to be expected that the adjusted table will be better than the original one in respect of every entry. This method can be extended, when a series of tables are available, to allow for autocorrelated, systematic and proportional errors.

5. FURTHER USES FOR THE ALL REGIONS ACCOUNTS

In section 3 above we considered an accounting system for a closed world of n regions in which there was never any difficulty of principle in assigning economic activities to regions. We must now remove these two restrictions since each is important in practice. Let us consider the two problems separately.

The introduction of an outside world into tables 3 or 4 presents no difficulty since it is equivalent to introducing an $(n+1)$th region the internal transactions of which are of no particular interest. Thus table 3 would be bordered to the right and below by a further set of three row and column pairs in respect of the rest of the world. The block in the bottom right-hand corner of this extended table would contain only zeros, except for a balancing entry as in table 5, and the remaining blocks would contain entries similar to the existing off-diagonal blocks of table 3. The derivation of a new table 4 from the new table 3 would proceed exactly as before.

The introduction of an outside world into tables 5, 6 and 7 can be performed in various ways. It is possible, as above, to introduce a set of accounts for the rest of the world and connect each region with them; or it is possible to introduce this new set of accounts and connect only the all regions accounts with them; or it is possible to combine the all regions accounts with those for the rest of the world. This last possibility would almost certainly be followed in an accounting system like that shown in table 7 since the object there was to reduce inter-regional information to a minimum.

But a set of accounts for all regions taken together can be used for quite a different purpose, namely to accommodate the transactions of a branch of economic activity which cannot really be assigned to regions at all. The obvious example of such a branch of activity is the central government which, though nominally located in the capital city, has branches and departments in all regions. These branches and departments may be responsible for a very large volume of transactions but have little or no regional significance. If they were assigned to regions it is likely that considerable difficulty would be met in tracing all the inter-regional transactions to which they give rise. If, on the other hand, they were assigned to the region which contained the capital city, they would in many cases completely change the apparent character of that region and would make it difficult to compare other regions with it. In these circumstances it may be better to treat central government as an all regions affair in which case it may be introduced into the accounting system in the various ways which have already been described. Other activities, such as railways, which spread over many regions, may conveniently be treated in the same way.

6. A NUMERICAL EXAMPLE OF REGIONAL ACCOUNTING

An actual example will now be given of a set of regional accounts constructed on the above lines. These accounts relate to the twelve civil defence regions of the United Kingdom in 1948 and are based entirely on the unpublished work of Phyllis Deane to whom all credit for this detailed statistical investigation must be given.

The example is set out in table 10 (facing page 132). It follows the lines of table 7 but is not exactly the same, since there is a set of accounts for the central government and since the United Kingdom is not a closed economy so that the final set of accounts relates to all regions and the rest of the world. The statistical work was completed some years ago and the figures are related to the *Blue Book* on national income and expenditure for 1954 [176].

The entries in table 10 are slightly more numerous than those in table 7 in which such flows as direct and indirect taxes and capital transfers were left out in the interests of simplicity. These entries can be followed by concentrating on the rectangle containing nine rows and columns at the bottom right-hand corner of table 10. This part of the system is reproduced symbolically in table 9. In following the explanation of these symbols certain general points should be kept in mind.

First, the consuming and accumulating activity of the central government
10

TABLE 9   TYPICAL ENTRIES IN THE REGIONAL ACCOUNTING SYSTEM OF TABLE 10

0	C_{jj}	V_{jj}	0	C_{jg}	0	X_{jr}	0	0
Y_{dj}	0	0	0	G_{jg}	0	Y_{jr}	G_{jr}	0
0	S_{jj}	0	0	0	T_{jg}	0	0	0
0	0	0	E	C	0	0	0	0
I_{gj}	D_{gj}	0	C	D	$-S_{gg}$	Y_{gr}	0	0
0	0	B_{gj}	0	0	0	$-V_{gr}$	0	B_{gr}
0	0	0	0	C_{rg}	0	0	0	0
0	0	0	0	G_{rg}	0	M_{rr}	0	0
0	0	B_{rj}	0	0	T_{rg}	0	N_{rr}	0

is taken out of the regions and accounted for separately. The producing activities of the central government on the other hand are included with all other forms of production in the regions. Second, the armed forces are entirely removed from the regions and are entered in the accounts for the central government. Third, all regions and the rest of the world are given a complete set of accounts which are made to balance by the introduction of appropriate transfers.

In table 9 the first three rows and columns relate to the set of three accounts for a region. The subscript j relates to one of the regions, g to the central government and r to all regions and the rest of the world.

All the imports into a region are routed through production and so the incomings into the production account of a region, the first row in table 9, are:

C_{jj} = consumption by the normal residents and local authorities of region j.

V_{jj} = gross investment in region j in fixed assets and stocks including stock appreciation. The investment of the central government is excluded and shown separately, entry V_{gr}.

C_{jg} = sales by region j to the central government for consumption purposes.

X_{jr} = net exports by region j to all other regions and the rest of the world.

As in a national production account, these incomings are matched by outgoings, shown in the first column of table 9, which are:

Y_{dj} = gross domestic product at factor cost of region j. In order to match V_{jj} this entry also includes stock appreciation.

I_{gj} = indirect taxes less subsidies paid by region j. The net total of all such taxes is paid into the income and outlay account of the central government although in fact some of them, such as local rates, are paid to local authorities in the region. These are subsequently returned to the local authorities in the region in entry G_{jg}.

The second row of the table shows the incomings into the income and outlay account of region j. These are:

Y_{dj} = gross domestic product of region j.

G_{jg} = current transfers, gifts and grants, received by region j from the central government. These include social security benefits, government grants to local authorities and the indirect taxes less subsidies levied by local authorities which have been routed through the income and outlay account of the central government. The reason for this is that it provides clear figures for Y_{dj} and I_{gj}.

Y_{jr} = net receipts of factor income by region j from all other regions and the rest of the world.

G_{jr} = current transfers received by region j from all other regions and the rest of the world.

These incomings are matched by the outgoings shown in the second column of the table which are:

C_{jj} = consumption of region j.

S_{jj} = gross saving of region j. In this investigation no estimates were made of regional depreciation and so all figures of product, investment and saving are gross of depreciation and also of stock appreciation.

D_{gj} = direct taxes on income paid by region j to the central government.

The third row of the table shows the incomings into the capital transactions account of region j. These are:

S_{jj} = gross saving of region j.

T_{jg} = net capital transfers received by region j from the central government. The largest incoming item under this heading in 1948 was the payment of war damage compensation and this was offset by payments of death duties by the regions to the central government.

The matching outgoings shown in the third column of the table are:

V_{jj} = gross investment in region j.

B_{gj} = net borrowing by the central government from region j.

B_{rj} = net borrowing by all other regions and the rest of the world from region j.

These fourteen entries characterize each region and are arranged in a way which is familiar from national accounting. Not only are the entries for the most part interesting in themselves but they are so arranged that the total incomings (or outgoings) of the three types of account are also of interest. For the production account this total is the gross domestic product of the region at market prices; for the income and outlay account it is the gross 'national' income of the region plus net current transfers; and for the capital transactions account it is the gross addition to the wealth of the region in the form either of investment in fixed assets and stocks or of a net increase in claims on others.

Let us now go through the set of accounts for the central government. As has been said already all central government productive activity is in the regional production accounts. It might be expected therefore that the first row and column for the central government would contain only zeros. They have however been used to account for the services of the armed forces and also to accommodate the residual error which appears in the national figures of the *Blue Book*. Thus E denotes the residual error and C denotes the consumption out of the income of the armed forces. Their total income is equal to $C + D$ where D denotes their direct tax payments to the central government.

The income and outlay account of the central government is a normal account apart from the entry $C + D$ which appears in both the row and the column. The remaining incoming entries are:

$\sum_j I_{gj}$ = total indirect taxes less subsidies received from the regions.

$\sum_j D_{gj}$ = total direct taxes on income received from the regions.

$- S_{gg}$ = dis-saving of the central government. The reason for this treatment of government saving will be explained shortly.

Y_{gr} = net factor income payments received by the central government from all regions and the rest of the world. The incoming component of this entry is government income from property. However, national debt interest is treated as negative income from property and, as this item is far larger than the incoming component, the entry itself is negative.

The matching outgoings are:

$\sum_j C_{jg}$ = government current expenditure on goods and services in the regions.

$\sum_j G_{jg}$ = government current transfers to the regions.

C_{rg} = government current expenditure abroad on goods and services.

G_{rg} = government current transfers abroad net.

The incoming entries into the capital transactions account of the central government are as follows:

$\sum_j B_{gj}$ = total borrowing by the central government from the regions.

$-V_{gr}$ = gross disinvestment in fixed assets and stocks by the central government. Like government saving, this entry has also been displaced from its expected position.

B_{gr} = net borrowing by the central government from the rest of the world.

The matching outgoings are:

$\sum_{j} T_{jg}$ = net capital transfers paid to the regions by the central government.

$-S_{gg}$ = dis-saving of the central government.

T_{rg} = net capital transfers paid to the rest of the world by the central government.

By looking at the incoming entries into the capital transactions account for the central government we can see the reason for the displacement of saving and investment. These entries are government borrowing and government dis-investment and, therefore, in total the negative of the gross addition to the wealth of the central government, that is to say they correspond to the negative of the outgoing entries in the capital transactions accounts for the regions.

The entries in the last set of accounts, for all regions and the rest of the world, have been arranged for the convenience of the regional and government accounts. The incoming entry in the production account is:

C_{rg} = central government purchases from the rest of the world.

The matching outgoings are:

$\sum_{j} X_{jr}$ = net imports from all regions by all regions and the rest of the world.

$\sum_{j} Y_{jr}$ = net payments of factor income to all regions by all regions and the rest of the world.

Y_{gr} = net payments of factor income to the central government. This is equal to government income from property which is generated in the regions less national debt interest which is treated as negative income from property.

$-V_{gr}$ = gross dis-investment of the central government.

M_{rr} = balancing transfers to income and outlay account, that is the excess of foreign exports to Britain and factor income received from Britain over British exports and factor income paid abroad.

The incoming entries in the income and outlay account are:

G_{rg} = current transfers (net) received by the rest of the world from the central government.

M_{rr} = net receipts from Britain from the sale of goods and services and in the form of factor income.

The matching outgoings are:

$\sum_j G_{jr}$ = net current transfers received from the regions by the regions and the rest of the world, that is net receipts of personal remittances from Britain.

N_{rr} = balancing transfer to capital transactions account, that is the rest of the world's balance on current account with Britain.

The incoming entries in the capital transactions account are:

$\sum_j B_{rj}$ = net borrowing by the regions and the rest of the world from the regions, that is British private lending to the rest of the world.

T_{rg} = capital transfers to the rest of the world from the central government.

N_{rr} = the rest of the world's balance on current account with Britain.

The matching outgoing is:

B_{gr} = foreign lending to the central government.

From these accounts the familiar domestic and national totals can readily be obtained. For example the gross domestic product at factor cost including stock appreciation is

$$\sum_j Y_{dj} + C + D + E = 10399$$

This total is the sum of the domestic products of the regions plus the value of the output of the armed forces which is not allocated to regions plus the residual error. Similarly, household and local authority consumption is

$$\sum_j C_{jj} + C = 8936$$

and total consumption, including the central government, is

$$\sum_j C_{jj} + C + \sum_j C_{jg} + C_{rg} + D = 10169$$

The unexpected appearance of D in this total is due to the fact that the armed forces are treated as outside the regions. The regions sell to the central government an amount which excludes the value of the services of the armed forces but includes the value of their consumption. Similarly the direct taxes paid by the regions exclude the direct taxes paid by the armed forces. Since the value of the services of the armed forces is equal to their consumption plus their direct tax payments, the latter, D, must be added to other central government expenditure on goods and services.

7. PRICE AND QUANTITY COMPARISONS BETWEEN REGIONS

If the regions in which we are interested are geographical subdivisions of a single country it may be safe, at least as a first approximation, to ignore regional differences in price levels. But even in a country as small and economically integrated as Britain there are considerable regional differences in money incomes and even in wages and salaries [46] which can hardly be due to a failure to make a proper assignment to regions as might be the case with profits. Again the price series for individual states given by Hurwitz and Stallings in [182] show a certain amount of variation so that the dollar can hardly have had at all times exactly the same purchasing power in each state.

If the regions are countries then some means of comparing the purchasing power of different currencies is needed. It is well known that exchange rates are, in general, a poor guide. Better results can be obtained by constructing price and quantity index-numbers. The problem is formally similar to pair-wise comparison over time and follows the general structure of such comparisons as set out for example in [150]. In practice, however, inter-country comparisons are usually harder to make than comparisons over time because the differences in institutional arrangements, the commodities used and the information actually available and its classification are much greater between countries than they are between time periods that are not very distant from each other.

If we consider a pair of domestic product accounts which show on one side final product, that is consumption, investment and net exports and on the other side values added in different branches of production, then we can in principle construct consistent price and quantity index-numbers from either side of these accounts as shown in [150]. These methods have been used by the O.E.E.C. and the results obtained from them are set out in [62] and [119].

However, difficulties of consistency arise if more than two countries are to be compared. If we compare country 1 with country 0 and then country 0 with country 1, consistency requires that the product of the comparisons should be 1. With Fisher's ideal index-numbers, we have for quantities

$$\left[\frac{p_0' q_1}{p_0' q_0} . \frac{p_1' q_1}{p_1' q_0}\right]^{1/2} \left[\frac{p_1' q_0}{p_1' q_1} . \frac{p_0' q_0}{p_0' q_1}\right]^{1/2} \equiv 1 \tag{7}$$

But with three countries, 0, 1 and 2, we obtain:

$$\left[\frac{p_0' q_1}{p_0' q_0} . \frac{p_1' q_1}{p_1' q_0}\right]^{1/2} \left[\frac{p_1' q_2}{p_1' q_1} . \frac{p_2' q_2}{p_2' q_1}\right]^{1/2} \left[\frac{p_2' q_0}{p_2' q_2} . \frac{p_0' q_0}{p_0' q_2}\right]^{1/2} \equiv \left[\frac{p_0' q_1}{p_0' q_2} . \frac{p_1' q_2}{p_1' q_0} . \frac{p_2' q_0}{p_2' q_1}\right]^{1/2} \neq 1 \tag{8}$$

unless, the quantity vectors, q_0, q_1 and q_2, being arbitrarily different, the price vectors, p_0, p_1 and p_2 are identical.

In price comparisons the roles of the p's and q's are interchanged and this led to the adoption of the concept of an average European quantity structure in the comparisons between Europe and the United States given in [62]. By adopting this average structure for all European countries, comparisons can be made between European countries, which are completely consistent though they may not be satisfactory in other ways. Thus for example the price comparison between two small countries may be based on a quantity structure which bears little relationship to the actual quantity structures in either country.

These difficulties have been resolved by van Yzeren in [192]. He considers a vector, r, say, whose elements are the unknown numbers of currency units in different countries which have equivalent purchasing power. He then introduces a matrix A whose elements are functions of the prices and quantities in the different countries and considers the characteristic equation

$$(\lambda I - A) = 0 \tag{9}$$

from which he derives r, up to a factor of proportionality, as the characteristic vector of A which corresponds to the largest characteristic root, λ_1 say, that is, the vector which satisfies the equation

$$(\lambda_1 I - A)r = 0 \tag{10}$$

Three different methods are given for calculating the elements of A, which lead to slightly different results. The third however may be regarded as a synthesis of the first two and as an extension of Fisher's ideal index-number to any number of countries, in the sense that if the $n(n-1)/2$ Fisher index-numbers between all pairs of n countries happened to be consistent, the ratios of these numbers would be exactly reproduced by it.

This problem has also been considered by Geary who gives a simple form of solution in [61].

8. STRUCTURAL ANALYSIS OF REGIONS

Regional accounting data such as have just been described can be used for a number of purposes. In this section let us consider the position when the hypotheses we can frame are only of a vague and general character and later turn to cases where they are more specific.

Some regions may be economically alike yet for various reasons quite different from other regions. Thus we can apprehend, though we may not be able to formulate precisely, the concept of regional structure. The set of accounts presented in table 10 provide us with fourteen transactions eleven of which are independent for each of twelve regions. Can we use this information to throw some light on regional structure?

We might approach this problem by employing the concept of distance as defined in works on n-dimensional geometry, such as [145]. Suppose we have n regions, $j, k = 1, 2, \ldots, n$, and m independent transactions, $r, s = 1, 2, \ldots, m$. These transactions could be represented along m orthogonal axes and each region would then be represented as a point in m-dimensional space. The square of the distance of region j from the origin of this space is given by

$$d_{j0}^2 \equiv \sum_r \sum_s \delta_{rs} x_{rj} x_{sj} \equiv x_j' x_j \tag{11}$$

where $\delta_{rs} = 1$ if $r = s$ and otherwise zero, and where $x_j \equiv \{x_{1j}, \ldots, x_{mj}\}$. The square of the distance between two regions, j and k, is given by

$$d_{jk}^2 \equiv \sum_r \sum_s \delta_{rs}(x_{rj} - x_{rk})(x_{sj} - x_{sk})$$

$$\equiv (x_j - x_k)'(x_j - x_k) \tag{12}$$

Several points arise in connection with this measure. First, we should probably not wish to distinguish regions merely by their size. If population can be taken as a measure of size we could make the necessary adjustment by dividing the transactions for each region by that region's population. Second, we should probably not wish to allow the analysis to be dominated by large transactions. In this case we should improve matters by normalizing each transaction, that is by putting $\sum_j x_{rj}^2 = 1$ for each r. Finally, we should probably wish to take account of the fact that the different transactions are correlated.

If we denote the normalized per head figures by x_{rj}^* and the corresponding vector for region j by x_j^*, then we can transform the actual transactions into a set of hypothetical orthogonal transactions, with values z_j, say, for region j, by means of a linear transformation

$$x_j^* = A z_j \tag{13}$$

where A is a non-singular matrix of order m. Since the z's are orthogonal we can adjust for the correlations among the x^*'s by replacing (11) by

$$d_{j0}^2 \equiv z_j' z_j \equiv x_j^{*'}(AA')^{-1} x_j^*$$

$$= x_j^{*'} R^{-1} x_j^* \tag{14}$$

where R is the matrix of zero-order correlations between the transactions. In a similar way (12) becomes

$$d_{jk}^2 \equiv (x_j^* - x_k^*)' R^{-1} (x_j^* - x_k^*) \tag{15}$$

which expresses the square of the distance between two regions, j and k, as a quadratic form in the differences of the observed values of their transactions per head expressed in normalized form.

Within regions we can calculate $n(n-1)/2$ values of d_{jk}^2 and may then attempt to group the regions into clusters such that the average value of d^2 within clusters is small compared with the average value of d^2 between clusters. This problem is discussed by Rao in [132]. He points out that no formal rules can be given for finding clusters because a cluster is not a well-defined concept. He adopts, however, a simple device, due to Tocher, which proceeds as follows. Start with the two regions with the smallest d^2 and find a third region which has the smallest average d^2 from the first two. Then find a fourth region with the smallest average d^2 from the first three. Continue in this way until the next nearest region has what appears to be a high average value of d^2 from the regions already included in the cluster. Do not include this region in the cluster and start again with the regions excluded.

It will be noticed that the method just described has much in common with factor analysis as used by psychologists in the analysis of intelligence tests. If we assume m general factors then (13) gives the relationships between the factors, the z's, and the observations, the x's. The equation, $R = AA'$, of which use is made in (14), shows the correlation matrix, R, in terms of the pattern matrix, A, and is sometimes called the fundamental theorem of factor analysis.

An important feature of this type of analysis is that in practice it is frequently possible to account for a large part of the observed variation with a relatively small number of factors. In geometrical terms this is equivalent to the possibility of finding a unit f-sphere, where f is small compared with n (the number of regions), such that the point representing each transaction on the unit n-sphere is close to its f-factor representation on the unit f-sphere. An economic application of this technique can be found in [146].

9. PARTIAL ANALYSIS WITH REGIONAL DATA

Table 10 contains information which can be used to study many relationships in the economy. It provides in fact a form of cross-section data in which the units are regions rather than the more familiar households. As an example let us consider saving behaviour. This example is in fact little more than an exercise because business and personal saving in the regions were taken as fixed proportions of gross profits and personal income, but it shows how regional information can be used when the basic data are adequate.

In terms of table 10, for region j the gross saving is S_{jj} and the gross disposable income is $C_{jj} + S_{jj}$. In this case S_{jj} includes depreciation and stock appreciation in the region in addition to all saving proper by businesses, local authorities and households. If we take out the figures for saving and disposable income for the twelve regions and reduce them to a per head basis we see that there is a clear positive relationship between the two variables.

In 1948 most private saving was made by businesses and in addition the available data include depreciation and stock appreciation in saving. Consequently it might be supposed that a region whose domestic product was large in relation to its disposable income would show a high level of saving at any given level of disposable income.

The gross domestic product at factor cost of region j is Y_{dj}. If we put $y = $ gross saving, $x_1 = $ gross disposable income, $x_2 = $ gross domestic product at factor cost $\div$ gross disposable income, we might test the hypothesis that

$$y = \beta_0 + \beta_1 x_1 + \beta_2 x_2 + \epsilon \tag{16}$$

where the β's are constants and ϵ is a disturbance. If we fit this relationship to the twelve sets of regional observations we find that

$$y^* = -40{\cdot}7 + 0{\cdot}2043\,x_1 + 24{\cdot}54\,x_2$$

$$(6{\cdot}1) \quad (0{\cdot}0228) \quad (7{\cdot}18) \tag{17}$$

where y^* is the calculated value of y and the figures in brackets are standard errors. For this equation the standard error of estimate is $1{\cdot}8$ and the proportion of the observed variance accounted for is $0{\cdot}995$. In this analysis each region is treated as a single observation and no attempt has been made to weight them in any way.

In this relationship all the constants are significant, and x_2 serves quite well to shift each region on to a single line connecting saving and disposable income. But while x_2 varies from region to region its value is presumably approximately constant for any particular region. On this basis the marginal propensity to save on the above definitions is about one-fifth. The regional average in 1948 was approximately one-eighth. Thus it would appear that the marginal propensity is higher than the average but not to the extent that is sometimes suggested in analyses based on time series.

As already explained, the assumptions made in estimating regional saving and the lack of direct information on transfers of factor income and gifts between regions which made these assumptions necessary, result in numerical estimates, as in (17), which are purely illustrative. This would not, of course, be the case if these deficiencies could be removed. Even then, however, it would only be possible to test dynamic saving hypotheses, as for example in [163], if the regional accounts were available for a series of years.

Regional analysis in which the regions are countries may also be illuminating, particularly if a study of variation over time can be combined with a study of variation over space. Examples of the use of a cross-section of countries for purposes of demand analysis are given by Juréen in [77] and by Gilbert and associates in [62].

10. NATIONAL ACCOUNTING MODELS WITH REGIONAL DATA

In [149], a simple transaction model, in which the individual outgoings in each account are assumed to be determined by the total incomings of the account, is based on a national accounts matrix rather than an input–output matrix. In this case it is safe to say that the assumption of simple proportionality between inputs and the corresponding outputs is unsatisfactory; even an assumption of linearity is extremely bold but nevertheless, in the example cited, gave quite interesting results.

In principle a similar model can be applied to a regional matrix. The rejection of the assumption of proportionality in favour of linearity means, as always happens in such cases, that more information is required. Thus it would be necessary to have a time series of regional accounts. It would also be necessary to recognize that the estimation of the parameters would involve a problem of statistical specification which is not present when input–output coefficients are calculated from a single matrix. This estimation problem has been studied by Briggs in [25, 26].

The model is similar in form to the input–output model. Thus if y is a vector of total revenues into each endogenous account, $\hat{y}$ is a diagonal matrix constructed from y, i is the unit vector, x is a vector of exogenous outgoings, Y is a matrix of endogenous flows and M and N are matrices of constants, then the accounting requirements of the model are

$$y \equiv Yi + x \tag{18}$$

that is the total incomings of each account are the sum of endogenous and exogenous incomings. The assumption of the model is

$$Y = M + N\hat{y} \tag{19}$$

that is each endogenous outgoing from an account is a linear function of that account's total of incomings. By combining (18) and (19) we obtain

$$y = Mi + Ny + x$$
$$= (I - N)^{-1}(m + x) \tag{20}$$

where I denotes the unit matrix and $m \equiv Mi$. This equation gives y in terms of x. However there is no difficulty in finding certain y's or x's in terms of a mixture of y's and x's. For example in a system with three endogenous subsystems, we might wish to know y_1 in terms of x_1, x_2 and y_3. If we write the first line of (20) in the partitioned form

$$\begin{bmatrix} y_1 \\ \cdots \\ y_2 \\ \cdots \\ y_3 \end{bmatrix} = \begin{bmatrix} m_1 \\ \cdots \\ m_2 \\ \cdots \\ m_3 \end{bmatrix} + \begin{bmatrix} N_{11} & N_{12} & N_{13} \\ \cdots & \cdots & \cdots \\ N_{21} & N_{22} & N_{23} \\ \cdots & \cdots & \cdots \\ N_{31} & N_{32} & N_{33} \end{bmatrix} \begin{bmatrix} y_1 \\ \cdots \\ y_2 \\ \cdots \\ y_3 \end{bmatrix} + \begin{bmatrix} x_1 \\ \cdots \\ x_2 \\ \cdots \\ x_3 \end{bmatrix} \tag{21}$$

then we see that

$$y_1 = [I - N_{11} - N_{12}(I - N_{22})^{-1}N_{21}]^{-1}\{[m_1 + N_{12}(I - N_{22})^{-1}m_2]$$
$$+ [N_{12}(I - N_{22})^{-1}N_{23} + N_{13}]y_3 + [x_1 + N_{12}(I - N_{22})^{-1}x_2]\} \tag{22}$$

Since nothing has been said about the ordering of the accounts, it can be seen that this approach allows for considerable flexibility in the choice of knowns and unknowns.

11. EXTENDED ACCOUNTING MODELS: REGIONAL INPUT–OUTPUT

In a system of regional accounts, like the one set out in table 10 of section 6 above, it is always possible in theory to subdivide the accounts and so produce a more informative accounting structure. If each of the regional production accounts is subdivided by industry, there arises a system of regional input–output tables each set in its own national accounting framework. If the regional accounts relating to consumption and accumulation are removed from the regions and consolidated, then there arises a system of regional input–output tables each connected to one or more non-production accounts for all regions. In this way the familiar input–output matrix is extended so as to reflect the locative as well as the industrial aspect of production and the corresponding model describes inter-regional trade as well as regional production.

In the original model of this kind proposed by Isard in [74], the United States is divided into three regions and the production in each region is divided into twenty industries. In this model each industry in each region is regarded as a separate industry so that the economy consists of 60 industries rather than 20. The coefficients in this system are assumed to be fixed so that if industry r in region j requires input from industry s it must get it in fixed proportions from each region in which industry s is located. This assumption combined with a knowledge of the final demands on each industry in each region enables all output levels in all regions to be determined. If the regional sources of final demand are known then the trade levels and balances as well as the output levels can be calculated.

Since the appearance of Isard's original model, two main variants have been developed which we will now consider.

12. A CENTRALLY CONNECTED REGIONAL MODEL

In [95], Leontief proposed a model of a somewhat different kind. In this it is assumed, first, that commodities can be classified as either local or national and that the former are not subject to inter-regional trade but are balanced at the local level; second, that the industries in the different regions which produce a commodity which is subject to inter-regional trade, contribute to the total output of that commodity in fixed proportions; and third, that a given industry has the same cost structure in each region.

With this model output levels are determined if final demand for local commodities is known in each region and if final demand for national commodities is known for all regions together. If final demand for national commodities can be subdivided by region, then a trading pattern as well as an output pattern will emerge.

This model can be set out as follows. Let local commodities be denoted by the subscript r and national commodities by the subscript s. Then for the economy as a whole

$$
\begin{bmatrix} q_r \\ \cdots \\ q_s \end{bmatrix} = \begin{bmatrix} A_{rr} \vdots A_{rs} \\ \cdots \vdots \cdots \\ A_{sr} \vdots A_{ss} \end{bmatrix} \begin{bmatrix} q_r \\ \cdots \\ q_s \end{bmatrix} + \begin{bmatrix} e_r \\ \cdots \\ e_s \end{bmatrix}
$$

$$
= \begin{bmatrix} D_{rr} \vdots D_{rs} \\ \cdots \vdots \cdots \\ D_{sr} \vdots D_{ss} \end{bmatrix} \begin{bmatrix} e_r \\ \cdots \\ e_s \end{bmatrix} \tag{23}
$$

where the q's are output vectors, the e's are final demand vectors and the A's are input–output coefficient matrices. For convenience, D is written in place of $(I-A)^{-1}$. Let a region be denoted by the subscript j placed in front of a vector so that, for example, $_jq_s$ denotes the vector of output levels in region j of commodities entering into inter-regional trade, and let the fixed proportions in which region j contributes to the total output of such commodities form the elements of a vector $_jb_s$. Then for region j

$$
_jq_s = {_jb_s}q_s
$$
$$
= {_jb_s}(D_{sr}e_r + D_{ss}e_s) \tag{24}
$$

from (23). Also

$$
_jq_r = (I-A_{rr})^{-1}(A_{rs}{_jq_s} + {_je_r})
$$
$$
= (I-A_{rr})^{-1}[A_{rs}{_jb_s}(D_{sr}e_r + D_{ss}e_s) + {_je_r}] \tag{25}
$$

since, for locally balanced commodities, (23) holds for each region.

The trading pattern which emerges from this model shows each region's vector of imports and each region's vector of exports, but it does not show either the source of the imports or the destination of the exports. In other words, the regions are not connected directly but only indirectly through the economy as a whole.

Generalizations of this model allowing for a hierarchy of regions and for simple dynamic elements are given in [95].

13. VARIABLE COST STRUCTURES AND BILATERALLY FIXED TRADING PATTERNS

A complementary model has been given by Moses in [113]. Consider an economy with n industries and m regions, and let typical industries be r and s and typical regions be j and k. Suppose each region has a technical structure given by $_jA$ of order n and suppose that each commodity has a trade structure given by T_r of order m. The element $_ja_{rs}$ of $_jA$ indicates the amount of the product of industry s required per unit of output by industry r in region j. The element $_{jk}t_r$ of T_r indicates the proportion of region k's purchases of commodity r which comes from region j.

The information contained in the m technical matrices, $_jA$, and the n trading matrices, T_r, can be arranged to form two matrices which provide the basis of the model. Consider a matrix, A, which contains the $_jA$ as diagonal submatrices: for example, with three regions,

$$A = \begin{bmatrix} {}_1A & 0 & 0 \\ 0 & {}_2A & 0 \\ 0 & 0 & {}_3A \end{bmatrix} \tag{26}$$

A second matrix, T say, can be formed with the T_r as diagonal submatrices. With two commodities

$$T = \begin{bmatrix} T_1 & 0 \\ 0 & T_2 \end{bmatrix} \tag{27}$$

Both A and T are of order mn. From T a matrix T^*, in which the elements of T are rearranged in a convenient way, can be formed by an orthogonal transformation. Thus

$$\begin{aligned} T^* &= UTU^{-1} \\ &= \begin{bmatrix} {}_{11}\hat{t} & {}_{12}\hat{t} & {}_{13}\hat{t} \\ {}_{21}\hat{t} & {}_{22}\hat{t} & {}_{23}\hat{t} \\ {}_{31}\hat{t} & {}_{32}\hat{t} & {}_{33}\hat{t} \end{bmatrix} \end{aligned} \tag{28}$$

in the present example, where U is the orthogonal matrix

$$U = \begin{bmatrix} 1 & 0 & 0 & 0 & 0 & 0 \\ 0 & 0 & 1 & 0 & 0 & 0 \\ 0 & 0 & 0 & 0 & 1 & 0 \\ 0 & 1 & 0 & 0 & 0 & 0 \\ 0 & 0 & 0 & 1 & 0 & 0 \\ 0 & 0 & 0 & 0 & 0 & 1 \end{bmatrix} \tag{29}$$

In view of the definitions of T and U, $T'i = T^{*\prime}i = i$. Each $_{jk}\hat{t}$ in (28) has diagonal elements $\hat{t}_{jk1}, \ldots, {}_{jk}\hat{t}_n$.

The m regional output vectors, $_jq$, and the m regional vectors of final demand, $_je$, may be written down in order one below the other to yield vectors of type $mn \times 1$ which in the case of three regions are

$$q = \{ {}_1q_1, \ldots, {}_1q_n \, {}_2q_1, \ldots, {}_2q_n \, {}_3q_1, \ldots, {}_3q_n \}$$

and

$$e = \{ {}_1e_1, \ldots, {}_1e_n \, {}_2e_1, \ldots, {}_2e_n \, {}_3e_1, \ldots, {}_3e_n \}$$

If the regions did not trade at all, the relationships between their final demands and output levels could be written in the usual form, $q = Aq + e$, which contains m independent sets of equations, one for each region. With the introduction of trade, as represented by (28), this system becomes

$$q = T^*(Aq+e)$$
$$= (I - T^*A)^{-1}T^*e \qquad (30)$$

which yields a trading pattern connecting the regions pairwise in the case of each commodity.

This model has fixed trading patterns for all goods but no restrictions on relative regional outputs. It thus contrasts with the model of the preceding section which allows: variable relative regional outputs but no trade in locally balanced goods; and variable trading patterns but fixed relative regional outputs for goods entering into regional trade.

The complementarity between the two models can be seen by an example. Model (30) might show that in some cases regional output requirements exceed regional capacities although for the economy as a whole capacities are sufficient to meet demand. The model of (24) and (25) might then be used to show how the trading pattern of goods which are balanced nationally might be changed to meet the new demands.

A dynamic version of this model is given by Fei and Moses in [113].

14. LOCAL AND REGIONAL INPUT–OUTPUT STUDIES

At the regional level it is the inter-industry aspect of social accounting that seems to have been most highly developed. Brief references to a number of these studies are made in this section. It will be seen that for the most part they relate to a particular area of a country rather than to a system of accounts covering all the regions of a country.

Belgium. Three tables, all at 1953 prices, with 11 industrial sectors relating to the Liège region are given by Derwa in [49]. The same author gives a table with 25 sectors for 1953 in [48].

France. A table with 16 industrial sectors relating to Lorraine in 1952 is given by Bauchet in [15].

Holland. A table with 28 industrial sectors relating to Amsterdam in 1948 is given in [116].

Italy. An interesting application of inter-regional input–output analysis is given by Chenery in [40, 42]. The model relates to Northern and Southern Italy and has many features in common with the one set out for the United States by Moses in [113]. Households are included among the intermediate sectors but different marginal propensities are calculated for each region; trade coefficients are termed supply coefficients. The main difference is that instead of forming one large matrix of technological and trading coefficients, an iterative method is followed so that the solution is built up by successive approximation. The importance of this is that it enables the investigator to see when capacity and other limitations are being encountered and to change some of the coefficients to take account of this fact. This investigation provides a very good example of the use of a comparatively simple model to reflect complexities in the real world which cannot easily be built into it but which can be introduced by using an appropriate sequence of calculations coupled with a changing of certain coefficients where the preceding results show this to be necessary.

Japan. Tables with 182 industrial sectors for 1951 and 36 sectors for 1951 and 1954 at 1951 prices relating to the Kinki area and the rest of Japan are given in [76].

Sweden. A table with 62 industrial sectors relating to Stockholm in 1950 is given by Artle in [8]. Though based primarily on input–output concepts, this study describes and applies many methods of regional analysis.

United States. A table with 26 industrial sectors relating to the state of Utah in 1947 is given by Moore and Petersen in [108]. Several applications of these data are made including an interesting study of local income inter-actions and multipliers.

In [113] Moses gives an inter-regional input–output table for the United States and uses it in conjunction with the model set out in section 13 above. The table relates to 1947, and in it the economy of the United States is divided into three regions which may be called East, Central and West. Each region is subdivided into 11 industries of which one is households. The table thus contains a standard production coefficient matrix based on the national study for 1947 and trade coefficient matrices based on a number of sources which are discussed in [112]. The model is used for two purposes. First, to show the effect on output levels and trade balances of a hypo-thetical increase of 10 per cent in each element of final demand in the East and, second, to examine the stability of trading patterns. Trade coefficients are given for 1947, 1948 and 1949, and inter-regional shipments in 1947 are estimated by applying 1949 trade coefficients to base-period requirements. Although there is some year-to-year variation in trade coefficients they show, on the whole, considerable stability. In the 1947–1949 exercise just mentioned, the calculated shipments from each region differed from the actual values by at most 2 per cent though calculated shipments to individual regions differed in some cases by as much as 7 per cent from the actual values.

Another interesting regional model, relating to the United States in 1949, is given by Heady and Candler in [69]. In this case special emphasis is placed on agriculture which is divided into a crop sector and a livestock sector. For each of these sectors, information is given for six areas among which the country is divided. In addition there are six industrial sectors without

11

regional subdivision, and foreign trade and government are treated as intermediate sectors. The matrix multiplier for this system of 20 sectors is given.

An inevitable difficulty with input–output models is that they are likely to suffer from the rigidity of their assumptions though, as indicated in [40] some steps, short of a reformulation of the model in programming terms, can be taken by adopting an iterative method of solution coupled with a change in coefficients as various constraints are encountered. Alternatively, a model may be adopted which contains all the elements, demand and supply functions and transport costs, which are theoretically desirable in solving interregional pricing and trade problems at the industry level. Interesting applications of this kind of model to the livestock-feed economy of the United States are given by Fox in [58].

In concluding this section it is convenient to refer to two matters which do not fit into the regional classification used in describing data and models.

The first is the use of input–output data for international comparisons of the structure of production. In [43] Chenery and Watanabe have compared the productive structures of Italy, Japan, Norway and the United States and in [38] Cao-Pinna has compared Italy and Spain.

Second, input–output tables are usually compiled for countries and occasionally for smaller regions. In one case a table has been prepared for a larger region, namely the member countries of the O.E.E.C. A tentative table with 27 industrial sectors relating to 1953, together with the matrix multiplier and some illustrative examples, is given by Kirschen and his associates in [84].

15. FURTHER DEVELOPMENTS AND CONCLUSIONS

In this paper I have tried to survey the problems of adapting the ideas of social accounting to provide materials and methods for regional analysis. I have tried to keep statistical methods and econometric models in step with the information available and with the hypotheses in which the investigator is interested. At each stage in the development of basic data there are certain questions that can be answered and certain methods and models that are appropriate.

This survey is certainly incomplete but even so it has ranged over a considerable variety of topics. However, as indicated at the end of the preceding section, it has stopped short at a point where some readers might think that the discussion was just about to become interesting. The most highly developed model considered is the input–output model. Models of this type are designed to work out the approximate consequences of a rigid system of relationships. As we have seen, by adopting an iterative method of computation, something can be done to reduce the inherent rigidities of these models, but they are not well adapted to handle situations involving choice, that is to say situations in which something is to be maximized or minimized. For this purpose we must go a stage further and formulate the models in programming terms. In a sense, for the static case, it may be said that this is done implicitly in the model of spatial-equilibrium referred to in [58] since the Lagrange-multiplier technique of maximizing a function subject to constraints lies at the bottom of conventional supply and demand analysis. But in simple transaction models, whether of the national accounting or of the input–output type, we buy the possibility of answering complicated questions at the price of building in certain rigidities. It is the purpose of programming models to overcome these limitations by allowing problems of choice to enter again into the picture in such a way as to make calculation possible. In their static form, programming models are concerned with problems which in classical mathematics are solved by the method of undetermined multipliers; in their dynamic form, such models provide a means of solving computationally problems which in classical mathematics would be handled by the calculus of variations. The close connection between dynamic programming and this calculus is amply exemplified by Bellman in [16]. Since the problems we have to face in the real world typically involve constrained maxima over time, it is clearly desirable that the approach to regional problems which I have tried to describe should be carried through to its logical conclusion. However, the step is a big one and I cannot do more here than indicate some of the relevant literature.

Linear programming and activity analysis have developed from Hitchcock's paper [71], published in 1941, which gives a solution of the now famous locative problem of the most efficient routing of a product from several sources to several destinations. The first comprehensive work on activity analysis, namely [88], contains a solution to the extended problem of the efficient flow of world shipping. Programming methods have been applied successfully to many other locative problems and numerical examples of many of these have been brought together by Vajda in [184]. The immense development of this literature in recent years can be followed mainly in journals devoted to operations research and management science and in such comprehensive, annotated bibliographies as [14, 134]. A programming model for regional input–output is proposed by Moses in [114] and a study of allocation in space which links up with the general approach of this paper is given by Lefeber in [92]. A comprehensive introduction to inter-industry economics which covers both regional and programming aspects is given by Chenery and Clark in [41].

Thus plenty of material is available to carry the account begun in this paper through to its conclusion in programming models. Much of this material is concerned with the problems of small units such as firms or industries rather than with those of regions or nations and so would need adaptation. It would also be desirable to keep in mind other recent approaches to locative problems, such as the industrial complex analysis of

Isard, Schooler and Vietorisz [75], the computer methods for the control of interconnected systems of Kirchmayer [83] and the space-potential methods of Warntz [187].

If we look back over the developments outlined in this paper we see a picture of rapid and energetic growth in many parts of the world. The works cited in this paper number fifty-five but these form only a small part of the literature relevant to the subject. The extent of contemporary writing in this field and the fact that it takes place in almost every country raises two general problems: one of communication and one of standardization.

The first is perhaps the easiest to deal with. The methods and models suggested above give rise to systems of linear equations. The form of these systems is therefore simple but their size and in some cases the diversity of their components makes them extremely cumbersome both to write out and to take in unless a suitable notation is employed. The appropriate notation is undoubtedly that of matrix algebra which is extremely compact and ideally suited to linear operations. Its compactness has the further advantage that it brings out the similarities of different problems and so helps the transfer of known techniques from one area to another. A working knowledge of this notation should be regarded as essential by all students of the kind of topic discussed in this paper.

The second problem, namely standardization, is much more difficult. As in the case of national accounts, input–output accounts stand in great need of some international agreement on concepts, classifications and arrangements, particularly for the kind of comparative purposes with which we have been concerned here. An outline of the main problems involved is given in [162] and a more detailed treatment is given in [151].

APPENDIX

BRIEF NOTES ON THE CONSTRUCTION OF TABLE 10

These estimates were made at the Department of Applied Economics of the University of Cambridge during the years 1950–1954 in the course of an inquiry into the regional social accounts of the United Kingdom. For the final presentation of the accounts the national totals published in the *Blue Book* [176] for 1954 were taken as correct, and wherever the basic estimates were formed by building up from the regional level these were adjusted accordingly.

C_{ij} Household expenditure was estimated on the basis of a regional breakdown of the totals for each item of personal expenditure given in the *Blue Book*: the major source of indicators of regional expenditure was the 1950 Census of Distribution. Local authority expenditure was calculated from the returns submitted to the Ministry of Housing and Local Government.

V_j Gross regional investment was calculated from aggregates given in the Census of Production and *Blue Book* by assuming that capital investment in each industry was a constant fraction of its size (measured by employment) throughout the United Kingdom.

C_{jg} Except in relation to Northern Ireland and Scotland for which separate estimates were derived from the relevant Appropriation Accounts, central government purchases were split between regions largely on the basis of the distribution of government employees and their wages and salaries.

X_{jr} This is the balancing item in the production account.

Y_{dj} Incomes from employment were estimated by applying to numbers employed, data on average earnings in each region derived from a variety of sources of which the most important were the Censuses of Production and Distribution and the earnings enquiries of the Ministries of Labour and Agriculture. Gross domestic operating surplus was allocated to regions on the basis of gross true incomes assessed to Schedule D in 1949–50.

I_{gj} Local authority indirect taxes were derived from their returns to the Ministry of Housing and Local Government. Central government indirect taxes were estimated by applying regional indicators derived from a variety of sources to specific categories of tax receipts.

G_{jg} Grants to and from local authorities were derived from their returns to the Ministry. The distribution of other grants and gifts by regions was calculated largely on the basis of regional information published in government reports.

Y_{jr} This is the balancing item in the consumption account.

G_{jr} This was the residual which arose after calculating the regional receipts and expenditures of persons in an accounting system which was larger than the one reproduced here.

S_{jj} Undistributed profits were calculated as a percentage of the gross operating surplus of each region, and personal saving as a proportion of incomes other than transfer incomes. Local authorities' saving was calculated from their returns to the Ministry of Housing and Local Government.

D_{gj} The estimates for direct taxes are largely based on regional analysis of taxable income published in the relevant annual reports of the Inland Revenue Commissioners. Regional national insurance contributions were calculated by applying the appropriate rates to figures of the insured population.

T_{jg} Capital grants to local authorities were derived from the local authorities' returns. Except in the case of Scotland and Northern Ireland for which specific information was usually available, capital transfers to the private sector were estimated on the basis of a variety of information directly or indirectly analogous to the different types of transfer: thus, war damage compensation was allocated by analogy with the distribution of the rateable values of urban areas; refunds of excess profits tax by analogy with Schedule D incomes assessed; and war gratuities, etc., by analogy with the numbers of ex-members of the forces on release leave.

B_{gj} Net lending by central government to local authorities was calculated from the local authorities' returns. For the rest a national distribution by regions was made on the basis of total assessed incomes.

B_{rj} This is the balancing item on rest of the world account which was itself largely composed of residuals.

XI

A COMPARISON OF THE ECONOMIC STRUCTURE OF REGIONS BASED ON THE CONCEPT OF 'DISTANCE'

1. INTRODUCTION

This paper is concerned with comparing the economic structure of regions on the assumption that we cannot define the concept of economic structure in concrete terms. Those readers who believe that this stage of regional analysis has already been passed can have no interest in what is proposed here and would be wasting their time in reading this paper: they begin where it leaves off, or beyond.

In the development and expansion of his great work *The Vectors of Mind*, Thurstone [170] writes:

> The exploratory nature of factor analysis is often not understood. Factor analysis has its principal usefulness at the border line of science. It is naturally superseded by rational formulations in terms of the science involved. Factor analysis is useful, especially in those domains where basic and fruitful concepts are essentially lacking and where crucial experiments have been difficult to conceive. The new methods have a humble role. They enable us to make only the crudest first map of a new domain. But if we have scientific intuition and sufficient ingenuity, the rough factorial map of a new domain will enable us to proceed beyond the exploratory factorial stage to the more direct forms of psychological experimentation in the laboratory.

This paper is based on the concept of distance and the methods suggested for exploring regional structure are closely related to factor analysis. It is offered in the spirit of Thurstone's remarks where the 'rational formulations in terms of the science involved' either have not been made or are manifestly unsatisfactory.

The plan of this paper is as follows: section 2 is devoted to describing the problem of exploratory regional comparisons; section 3 introduces the data used; sections 4 and 5 outline the formal theory; section 6 sets out an application of the theory to the British economy; and section 7 contains some brief conclusions.

152

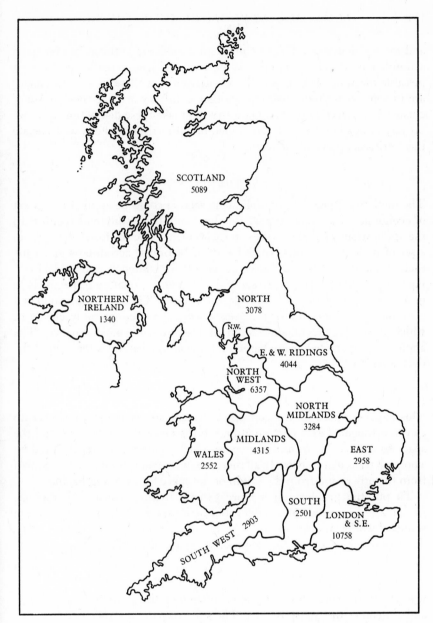

FIG. 1: The 12 Regions of the United Kingdom. (1948 Population in '000's)

2. THE PROBLEM

If we divide an economy into regions it is easy to see that these regions have a number of characteristics which are not the same in every case. Some regions are rich, others poor; some are agricultural, others industrial; some are workplaces, others dormitories; some are exporters, others importers;

and so on indefinitely. The first problem considered in this paper is: given a number of variables which may be used to characterize a region, is it possible to group the regions into clusters so that those in a given cluster are in some sense more like one another than they are like those in other clusters? The second problem is: if we can make these distinctions, can we say anything about the criteria implied in our analysis and can we translate these criteria into concrete terms?

3. THE DATA

The method proposed in the following section could be applied to any set of economic measurements available for a number of different regions. In the applications given here, these measurements take the form of entries in a set of regional accounts. In [X] a set of 'national' accounts is given for each of the twelve civil defence regions of the United Kingdom in 1948. Each region is provided with a set of three accounts, relating respectively to production, consumption and accumulation. These are consolidated accounts and contain in all fourteen distinct transactions, only eleven of which are independent since each of the accounts must balance. The applications given below are based on these eleven variables for each of the twelve regions.

4. THE METHOD

This paper is based on the concept of Euclidean distance. We can set up eleven orthogonal axes corresponding to the eleven variables, and call the space so defined the transaction space. Each region can be represented by a point in transaction space. In all there are twelve such points each separated from the others and from the origin of the space by a certain distance.

To put the problem on a more general footing, consider n regions, $j, k = 1, 2, \ldots, n$, and m independent transactions, $r, s = 1, 2, \ldots, m$. Then from standard works on n-dimensional geometry we know that the square of the distance of region j from the origin of the space, d_{j0}^2 say, is given by

$$d_{j0}^2 \equiv \sum_r x_{rj}^2 \equiv x_j' x_j \tag{1}$$

where x_{rj} is the value of transaction r in region j, the vector $x_j \equiv \{x_{1j}, \ldots, x_{mj}\}$ and x_j' denotes the transpose of x_j. The square of the distance between two regions, j and k, is given by

$$d_{jk}^2 \equiv \sum_r (x_{rj} - x_{rk})^2$$
$$\equiv (x_j - x_k)' \ (x_j - x_k) \tag{2}$$

As they stand, these definitions of distance are unsatisfactory. First, if one region is much larger, in population for example, than the others, it is likely that most, if not all, of its transactions will also be much larger than their counterparts in other regions. Accordingly this large region will appear as very distant from the others simply because it is large. To avoid this result we must adjust the regions for size. One way to do this would be to divide the transactions of each region by that region's population. If we denote the population of region j by π_j, then we should replace the vector x by x_j^* where

$$x_j^* \equiv \pi_j^{-1} x_j \tag{3}$$

Evidently there is nothing compelling about the use of population in this way. In (3) regions will no longer be distinguished by size, as measured by population, but rich regions, with a high level of production per head, will still tend to be distant from poor ones. If we wished to remove this distinction we should have to divide the regional transactions not by the region's population but by, say, its gross domestic product.

In fact each transaction of a region need not necessarily be divided by the same magnitude. Thus we might divide the individual outgoings of each account by that account's total outgoings ($\equiv$ total incomings). If this method were adopted the variables to be analysed would be similar to those in a matrix of input–output coefficients.

The point that emerges from this discussion is that as we change the variable used to standardize the observations we change the characteristics of the regions which contribute to the distance between them.

But this is not the only respect in which the original definition of distance is unsatisfactory. For, second, it may not seem sensible to allow distance to depend on the relative size of transactions. For example domestic product per head is a large transaction capable of large absolute variation from region to region. By comparison, saving per head is a small transaction showing, typically, a much smaller variation. In defining distance we may not wish to give one transaction more weight than another. If this is so we should be led to normalizing the transactions by making their sum of squares over the regions equal to 1 in each case. Thus if x_{rj}^* is the standardized value of transaction r in region j, we might convert it into a normalized measure by putting

$$\sum (x_{rj}^*)^2 = 1 \tag{4}$$

so that the normalized measures, x_{rj}^{**} say, are given by

$$x_{rj}^{**} \equiv \frac{x_{rj}^*}{\sum\limits_{j} (x_{rj}^*)^2} \tag{5}$$

or, in vector form,

$$x_j^{**} \equiv \hat{q}^{-1} x_j^* \tag{6}$$

where the rth element of the diagonal matrix $\hat{q}$ is $\sum\limits_{j} x_{rj}^{*2}$.

Finally, there is a third consideration to be taken into account. Even if we standardize and normalize the observations we have still made no allowance for the fact that the different variables used to characterize the regions are likely to be quite highly correlated. If each pair of transactions was perfectly correlated over the regions we should gain nothing by looking at more than one, since each would tell the same story. We can allow for the inter-correlations of the variables by reducing them to a set of hypothetical orthogonal variables in terms of which distance can be measured by means of (1) and (2) since they can properly be referred to a set of orthogonal axes.

If the variables (the transactions) are linearly independent then they are connected to the hypothetical orthogonal transactions, z_{rj} say, by the linear transformation

$$x_j^{**} = Az_j \tag{7}$$

where A is a non-singular matrix of order m. If we now calculate distance in terms of the z's we see that

$$d_{j0}^2 \equiv z_j'z_j \equiv x_j^{**'}(AA')^{-1}x_j^{**}$$
$$\equiv x_j^{**'} R^{-1}x_j^{**} \tag{8}$$

where R is the matrix of zero-order correlations between the transactions. In a similar way (2) becomes

$$d_{jk}^2 \equiv (x_j^{**}-x_k^{**})' R^{-1}(x_j^{**}-x_k^{**}) \tag{9}$$

which expresses the square of the distance between two regions, j and k, as a quadratic form in the differences of the observed values of their transactions, each expressed in standard, normal form.

From the observations we can calculate the elements of R and of the x_j^{**}. Accordingly from (9) we can calculate the $n(n-1)/2$ distances between the n regions. We may then attempt to group the regions into clusters in such a way that the average value of d^2 within clusters is small compared with the average value of d^2 between clusters. This problem is discussed by Rao in [132]. He points out that grouping is an arbitrary process, and that different definitions of the qualities distinguishing the members of one group from those of another will lead to differing procedures. This point will be taken up more fully in section 6 of this paper.

5. DISTANCES AND FACTORS

Equation (9) will be recognized as formally similar to Mahalanobis' generalized distance set out many years ago in [101 to 104]. In these applications, races, castes or tribes corresponded to regions and anthropometric measurements to transactions. An important difference which distinguishes these from the present application is that Mahalanobis had available a number of sets of measurements for each race, whereas in the regional analysis attempted here there is only a single set of measurements for each region.

Consequently Mahalanobis was able to obtain measures of the correlations between characteristics by pooling the correlations obtained within each racial group whereas in the present case the correlations are obtained by correlating pairs of transactions over regions. The statistical aspects of generalized distance have been discussed in detail by R. C. Bose in [20, 21], S. N. Bose in [23, 24], R. C. Bose and S. N. Roy in [22], and Roy in [136, 137].

Again, the matrix A in (7) will be recognized as the pattern matrix of factor analysis and the relationship $R = AA'$ of which use is made in (8) as what is sometimes called the fundamental theorem of factor analysis. In (7) the elements of z_j can be interpreted as a set of m orthogonal hypothetical transactions or factors in terms of which the actual transactions can be described. The matrix A contains in row s the weights necessary to construct transaction s as a linear combination of the hypothetical factors. In Hotelling's method of principal components [72], the first component or factor is chosen in such a way as to remove as large a part as possible of the sum of squares of the observations, that is to minimize the expression

$$\sum_r \sum_j (x_{rj} - a_{r1}f_{1j})^2$$

where f_{1j} is the value of the first factor in region j and a_{r1} is the element in row r and column 1 of A. The second factor is then chosen so as to remove as large a part as possible of the residual sum of squares left when the variation attributable to the first factor has been removed. Provided that the transactions are linearly independent this process, if continued, will eventually lead to m orthogonal factors or hypothetical transactions in terms of which the whole of the original sum of squares of the observations can be accounted for.

One of the interesting features of factor analysis is that, with correlated variables, it is usually possible to account for most of the original sum of squares with far fewer factors than variables. This situation can be described geometrically. Suppose there are three regions and any number of transactions. If the regions are represented by three rectangular axes and if the transactions are normalized, then each transaction can be represented as a point in the region space which is a unit distance from the centre of the co-ordinate system. Thus each transaction can be represented as a point on the unit sphere. If the third orthogonal factor accounts for a negligible amount of the total sum of squares, then the points representing the transactions will all lie close to a great circle drawn on the surface of the sphere. If both second and third orthogonal factors are unimportant then all the transactions will lie in the neighbourhood of a single point on the surface of the sphere.

Distance analysis and factor analysis are thus closely related: a distance analysis for regions is, in a sense, complementary to a factor analysis for transactions. But the roles of regions and transactions could be interchanged: we could imagine a factor analysis for regions and a complementary distance analysis for transactions.

An economic application of Hotelling's method is given in [146]; the variables were again transactions but the observations related to years rather than regions. The method described is, however, exactly the same.

6. APPLICATIONS

As mentioned above in section 2, the theory of the preceding section will now be applied to data for the economic activity of the United Kingdom during the year 1948. The twelve civil defence regions of the United Kingdom are shown with their population in 1948 in the sketch-map. The accounting data are based entirely on the unpublished work of Phyllis Deane and are described in more detail in [X]. In table 1 the actual figures for the eleven transactions used are set out. These are taken from a complete set of fourteen transactions, from which three are omitted corresponding to three accounting identities which hold between the fourteen transactions. The symbols given in table 1 are the same as those used in [X], and the data are given in millions of pounds sterling.

The analyses performed on the data of table 1 fall into three classes. In each analysis, the figures for each region have been divided by the regional population in order to remove a gross size effect, as in (3) above. As before, the variable x^*_{rj} denotes the value of the rth transaction in the jth region, and the vector x^*_j denotes the vector of all transactions for the jth region.

In analysis A the distances between each pair of regions, j and k say, are measured by (2), repeated here as

$$d^2_{jk}(A) \equiv (x^*_j - x^*_k)' \, (x^*_j - x^*_k) \tag{10}$$

Thus in this analysis the regions are represented by points in 11-dimensional space defined by eleven orthogonal axes corresponding to the transactions.

In analysis B the transactions have been first separately normalized *within* each region so that

$$d^2_{jk}(B) \equiv (x^{**}_j - x^{**}_k)' \, (x^{**}_j - x^{**}_k) \tag{11}$$

In this analysis each region is represented as a point on the unit hypersphere in 11-dimensional space defined again by eleven orthogonal axes corresponding to the transactions.

In analysis C the transactions have been normalized *over* regions, and account has been taken of the correlations between transactions so that

$$d^2_{jk}(C) \equiv (x^{**}_j - x^{**}_k)' \, R^{-1}(x^{**}_j - x^{**}_k) \tag{12}$$

TABLE 1   ACCOUNTING DATA USED IN THE ANALYSIS

Transaction No.	1	2	3	4	5	6	7	8	9	10	11
Transaction title	Consumption	Gross investment	Gross domestic product	Gross saving	Indirect taxes	Direct taxes	Central government borrowing	Sales to central government	Current transfers from central government	Capital transfers from central government	Current transfers from the rest of the world
Transaction symbol	C_{jj}	V_{jj}	Y_{dj}	S_{jj}	I_{gj}	D_{gj}	B_{gj}	C_{jg}	G_{jg}	T_{jg}	G_{jr}
1. North	482	106	520	51	68	95	-63	59	77	-1	0
2. E. and W. Ridings	690	171	844	105	103	167	-79	72	97	2	29
3. N. Midlands	551	127	620	72	85	126	-71	66	74	-2	7
4. East	498	91	489	65	88	90	-53	69	74	-3	-29
5. London and S.E.	2348	393	3096	440	499	566	-269	311	306	20	21
6. South	430	75	379	37	76	74	-56	108	68	-3	-32
7. South-west	451	89	465	56	71	88	-61	97	73	-3	-49
8. Wales	364	81	377	36	32	67	-45	44	71	0	-34
9. Midlands	719	179	908	103	139	178	-90	91	96	2	-23
10. North-west	1111	282	1300	153	186	253	-128	126	161	7	46
11. Scotland	878	173	947	113	97	169	-97	81	127	-11	44
12. N. Ireland	171	30	178	19	5	31	-16	14	26	0	-15

Note: The transactions omitted because of the adding-up constraints are: (i) net exports to the rest of the world $(X_{jr}) \equiv 3+5-1-2-8$; (ii) net factor income from the rest of the world $(Y_{jr}) \equiv 1+4+6-3-9-11$; (iii) borrowing by the rest of the world $(B_{rj}) \equiv 4+10-2-7$.

This is similar to the analysis A except that it is not assumed that the transactions lie on orthogonal axes. For a reason which will be discussed later, this analysis (C) has been carried out on the first three, the first six and the first nine transactions of table 1.

In each analysis the $(12 \times 11)/2 = 66$ elements of the distance matrix have been calculated, and attempts made to group the regions into some plausible classification. The purpose of grouping is clearly to place within the same group regions whose mutual distances are small, but as Rao [132] has pointed out, the exact method for grouping will depend on making the concept of a group more precise, as can be done in a number of arbitrary ways. Thus the method will differ, for example, according to whether simple clusters, rings or chains are looked for. The technique adopted is suitable for discovering simple clusters. First, two measures are defined, a measure of the compactness of a group of regions, and a measure of distance between groups of regions; these definitions are based on the idea of average distances. Let j and k be typical regions in group g, containing p regions, and let u and v be typical regions in group h, containing q regions. Then the measures of compactness are given by

$$D^2_{gg} \equiv \frac{2}{p(p-1)} \sum_{j<k} d^2_{jk} \tag{13}$$

$$D^2_{hh} \equiv \frac{2}{q(q-1)} \sum_{u<v} d^2_{uv} \tag{14}$$

and the distance between group g and group h by

$$D^2_{gh} \equiv \frac{1}{pq} \sum_{j=1}^{p} \sum_{u=1}^{q} d^2_{ju} \tag{15}$$

Beginning with all regions unclassified, the two with minimum distance are chosen; these form the nucleus of the first group with $D^2_{11} \equiv d^2_{jk}$, if j and k are these two regions. The remaining regions are then ordered according to their average distance from the regions in group 1. The region with the minimum average distance is then included in this group if its average distance is not greater than say αD^2_{11}, where α is an arbitrary constant. In this case D^2_{11} is recalculated for the three regions and the procedure repeated. Otherwise the third region forms the nucleus of a new group. The next nearest region to the third is included with it if its distance from it is less than αD^2_{22}, and the procedure repeated.

This procedure is one in which the criterion of inclusion changes through the analysis, as more regions are included in groups or new groups formed, and may be called 'grouping according to the sequential criterion'. Another method has also been tried in which the mean of all the elements of the distance table is first calculated, $\bar{d}^2$, say, and regions are classified together if $\bar{d}_{jk}^2$, say, is not greater than αd^2. This procedure may be called 'grouping according to the absolute criterion.'

The results for analysis A are given in tables 2 and 3. In table 2 the values of $d_{jk}(A)$ are given, and in table 3 the grouping, intra-group and inter-group distances are shown. The differences between the two methods are not very great, being mainly a matter of the demarcation line between groups II and III. In fact, it is possible to represent analysis A approximately on a line diagram, in which the regions are ordered as they appear in table 3, with London heading the list and Northern Ireland at the bottom. This is so because from table 3 it can be seen that

$$D_{gh} \propto D_{gi} + D_{ih} \tag{16}$$

for any g, h, i, provided the elements of the matrix are appropriately signed. The ordering arrived at is closely related to gross domestic product per person, and is therefore roughly equivalent to an ordering of the regions by the criterion of productiveness.

The results of analysis B are set out in tables 4 and 5. The normalization of the transactions within regions, that is the division of each transaction for a region by the sum of squares of all transactions for the same region, removes any factor which is consistently greater in one region than another over all transactions, and therefore places more emphasis on the structure of transactions, that is the size of the individual transactions in relation to each other. Once again the effect of using different grouping criteria is somewhat trivial, merely changing the demarcation line, as it were, between groups in a fairly simple way.

With regard to analysis C there are certain limitations which are characteristic of the present application. The method of analysis C is suggested, as described in section 5 above, by analogy with factor analysis, which was developed within the field of psychometrics. In intelligence tests there are usually no common units of measure between the variables, which are typically the scores of students when examined say in mathematics or in languages. Further, there are usually sufficient examination results to enable the correlation matrix between variables to be estimated reasonably well. In the present application to regional accounting we have, on the other hand, a common unit of account, in which the different transactions are measured, while the number of regions is very little greater than the number of transactions. If the number of regions is equal to (or less than) the number of transactions, the correlation matrix as estimated from the data has no inverse and $d_{jk}^2(C)$ cannot be calculated. The reason for this is that if the n points representing the regions are plotted in n dimensional space they define a figure which could be plotted in $(n-1)$ dimensional space, at most, from which it is not possible to estimate the parameters which could transform the n-space into a different n-space. If there is one

TABLE 2   INTER-REGIONAL DISTANCES (ANALYSIS A)

Regions	Regions											
	1	2	3	4	5	6	7	8	9	10	11	12
1. North	—	14·5	8·1	6·1	44·4	11·8	7·8	9·9	14·8	13·9	8·3	17·3
2. E. and W. Ridings		—	6·9	15·8	30·9	21·8	19·3	23·7	4·9	2·7	8·5	30·7
3. N. Midlands			—	9·3	36·7	15·5	12·9	17·4	7·7	6·3	4·5	24·8
4. East				—	43·5	8·3	6·3	11·9	15·2	14·7	9·9	19·5
5. London and S.E.					—	47·7	47·0	53·2	30·8	31·3	37·7	60·5
6. South						—	7·2	13·8	21·0	20·4	16·2	20·9
7. South-west							—	8·9	18·1	18·0	14·1	16·7
8. Wales								—	23·3	23·3	17·4	8·8
9. Midlands									—	5·4	10·8	30·6
10. North-west										—	7·9	30·7
11. Scotland											—	24·3
12. N. Ireland												—

TABLE 3   GROUPINGS OF REGIONS IN ANALYSIS A
BY ALTERNATIVE CRITERIA

Criterion 1.   Absolute criterion.

Regional groupings.   I. London and S.E.

II. E. and W. Ridings, N. Midlands, Midlands, North-west, Scotland.

III. North, East, South, South-west, Wales.

IV. N. Ireland.

Distances within and between groups.

	I	II	III	IV
I	0	34	46	61
II		7	17	28
III			10	17
IV				0

Criterion 2.   Sequential criterion.

Regional groupings.   I. London and S.E.

II. E. and W. Ridings, North-west, Midlands, Scotland.

III. North, N. Midlands, East, South, South-west, Wales.

IV. N. Ireland.

Distances within and between groups.

	I	II	III	IV
I	0	31	43	61
II		2	17	31
III			13	21
IV				0

Note: In the distance table the diagonal elements indicate the average distances between regions within a group, and the off-diagonal elements the distances between groups.

12

TABLE 4   INTER-REGIONAL DISTANCES (ANALYSIS B)

Regions	1	2	3	4	5	6	7	8	9	10	11	12
1. North	—	24·6	11·8	23·7	38·9	48·7	31·9	24·8	28·0	18·6	16·2	33·2
2. E. and W. Ridings		—	15·4	43·2	24·9	69·6	49·3	43·4	17·5	9·2	22·7	45·5
3. N. Midlands			—	29·6	31·3	55·6	37·0	33·0	20·3	10·2	17·0	38·5
4. East				—	49·2	33·5	21·5	26·9	41·0	37·2	33·2	36·7
5. London and S.E.					—	74·4	52·6	54·7	20·9	29·1	38·9	55·6
6. South						—	28·6	46·0	66·4	62·9	59·2	58·8
7. South-west							—	27·0	42·8	44·6	44·6	40·5
8. Wales								—	41·8	39·9	34·9	21·1
9. Midlands									—	19·2	33·9	47·0
10. North-west										—	19·9	45·1
11. Scotland											—	34·8
12. N. Ireland												—

TABLE 5   GROUPING OF REGIONS IN ANALYSIS B
BY ALTERNATIVE CRITERIA

Criterion 1.   Absolute criterion.

Regional groupings.   I. London and South-east.

II. Midlands, E. and W. Ridings, North-west, North, N. Midlands, Scotland.

III. East, South-west.

IV. Wales, N. Ireland.

V. South.

Distances within and between groups.

	I	II	III	IV	V
I	0	3·14	5·09	5·51	7·44
II		2·00	3·88	3·89	6·08
III			2·15	3·33	3·11
IV				2·11	5·28
V					0

Criterion 2.   Sequential criterion.

Regional groupings.   I. London and South-east, Midlands.

II. E. and W. Ridings, North-west.

III. North, N. Midlands, Scotland.

IV. East, South-west, Wales, N. Ireland.

V. South.

Distances within and between groups.

	I	II	III	IV	V
I	2·09	2·31	3·25	4·84	7·05
II		0·92	1·92	4·36	6·63
III			1·52	3·37	5·46
IV				2·99	4·34
V					0

TABLE 6 INTER-REGIONAL DISTANCES (ANALYSIS C): THREE TRANSACTIONS

Regions	1	2	3	4	5	6	7	8	9	10	11	12
1. North	—	13·7	7·7	15·2	36·9	25·5	7·1	7·2	15·3	17·2	10·1	26·4
2. N. and W. Ridings		—	8·9	25·6	36·3	34·6	20·2	18·4	6·0	9·4	18·3	35·6
3. N. Midlands			—	17·2	36·2	25·9	13·5	14·3	13·2	10·0	10·8	33·6
4. East				—	35·5	11·7	11·4	19·4	28·6	25·3	8·1	32·7
5. London and S.E.					—	43·4	35·5	41·4	35·9	41·0	30·9	43·8
6. South						—	22·5	29·3	38·6	31·5	18·8	42·8
7. South-west							—	8·9	21·2	23·4	8·9	23·0
8. Wales								—	18·3	22·8	16·1	21·2
9. Midlands									—	15·2	21·1	32·7
10. North-west										—	19·7	42·5
11. Scotland											—	30·9
12. N. Ireland												—

Regions

TABLE 7 GROUPINGS OF REGIONS IN ANALYSIS C
(THREE TRANSACTIONS, ABSOLUTE CRITERION)

Regional groupings. I. London and S.E.

II. E. and W. Ridings, North Midlands, Midlands, North-west.

III. North, Scotland, South-west, Wales.

IV. East, South.

V. N. Ireland.

Distances within and between groups.

	I	II	III	IV	V
I	0	37·4	36·4	39·6	43·8
II		10·9	17·8	29·1	36·3
III			10·1	19·9	25·6
IV				11·7	38·1
V					0

TABLE 8   INTER-REGIONAL DISTANCES (ANALYSIS C): SIX TRANSACTIONS

Regions	1	2	3	4	5	6	7	8	9	10	11	12
1. North	—	36·5	37·9	39·4	40·3	34·5	30·2	13·6	24·6	26·9	32·3	38·1
2. E. and W. Ridings		—	17·2	37·6	40·5	43·5	28·6	32·1	32·1	19·8	31·1	36·6
3. N. Midlands			—	42·2	42·8	32·6	29·6	36·3	33·8	29·8	36·5	35·8
4. East				—	44·5	43·1	19·1	36·3	42·8	33·2	39·6	44·4
5. London and S.E.					—	46·6	39·5	43·1	39·9	42·2	39·7	45·4
6. South						—	35·8	39·0	39·4	41·6	40·6	46·6
7. South-west							—	27·8	28·8	29·8	38·3	30·3
8. Wales								—	30·7	25·5	25·8	30·3
9. Midlands									—	29·6	47·0	41·6
10. North-west										—	27·9	45·4
11. Scotland											—	39·5
12. N. Ireland												—

TABLE 9   INTER-REGIONAL DISTANCES (ANALYSIS C): NINE TRANSACTIONS

Regions	Regions											
	1	2	3	4	5	6	7	8	9	10	11	12
1. North	—	44·9	39·7	43·6	44·6	44·3	43·9	36·9	31·0	38·6	35·9	45·1
2. E. and W. Ridings		—	35·1	44·3	41·4	44·2	41·5	37·9	37·7	23·0	40·4	40·4
3. N. Midlands			—	47·0	48·3	47·2	48·3	47·7	44·4	46·3	45·6	48·1
4. East				—	48·3	49·0	48·5	47·7	46·4	40·7	47·7	47·6
5. London and S.E.					—	48·5	49·0	47·1	43·9	45·3	45·7	48·9
6. South						—	48·7	47·6	45·8	41·8	47·2	48·1
7. South-west							—	47·6	44·8	44·7	46·4	48·7
8. Wales								—	48·1	40·7	48·5	46·3
9. Midlands									—	32·2	48·8	42·2
10. North-west										—	35·5	46·3
11. Scotland											—	44·4
12. N. Ireland												—

more region than the number of transactions the calculation is possible but extremely restricted. It is obvious that with 2 regions and 1 transaction the normalized distance between the regions always reduces to the same numerical value. And in the present application with 12 regions and 11 accounts the inter-regional distances will reduce to

$$d_{jk}^2(C) = 2400. \tag{17}$$

Accordingly the analysis has been applied with 3, 6 and 9 transactions, each taken in order from table 1. The distance matrix and groupings for the 3-transaction analysis are given in tables 6 and 7.

Tables 8 and 9 show the distance matrix for 6 and 9 transactions respectively. It will be noticed that in the 9-transactions table the elements are approaching $49 = \sqrt{2400}$ in each case.

7. CONCLUSIONS

In this paper several simple methods are presented for comparing the economic structure of a number of regions as revealed by their transactions in a given year.

From a technical point of view analysis A suffers because, though a gross size factor has been removed at the outset by expressing all variables per head of population, the results are likely to be dominated by any remaining tendency which causes all the transactions of one region to be systematically greater than those of another.

Analysis B, whilst being only slightly more difficult to compute than analysis A, removes this further factor by deflating the figures for each region by the sum of the squares of all transactions for the region. Since this is done separately for each region, problems of comparing the unit of account in different regions, as would arise in comparing different countries within the same geographical area, are avoided. Though the analysis does not formally take into account the possibility that the different transactions may be inter-correlated, this is largely revealed by the analysis, for it may be easily tested whether or not the results can be approximately shown on a line or plane diagram. Such analysis may help in the decision as to whether a formal factor analysis would be worth while.

Analysis C, which has a sounder theoretical base than the preceding analyses, leads to difficulties if the number of regions does not greatly exceed the number of transactions unless it is possible to replace the actual transactions by a relatively small number of hypothetical transactions or factors. To carry out the analysis it must be assumed that there is a common unit of account between regions, though not necessarily between transactions. Indeed any original unit of account between transactions is destroyed in the process of the analysis, which reduces all transactions to scales on which they show equal variability between regions. The number of calculations required in the analysis is very much greater than in analyses *A* and *B*.

In the present application the qualitative results of all the analyses are very similar. London and South-east region is usually clearly separated from the rest. Next come the regions which contain the other great conurbations and industrial centres of the United Kingdom, the East and West Ridings, North-west, Midlands, North-Midlands and Scotland. The East, South-west and Wales usually form a third group. Northern Ireland is usually clearly separated, and so also is the South region, which has no specific character, ranging from the South Coast, with Southampton, Portsmouth and holiday resorts, through sparsely populated agricultural counties, to as far north as Oxford.

XII

POPULATION MATHEMATICS, DEMAND ANALYSIS AND INVESTMENT PLANNING

1. INTRODUCTION

The main problem considered in this paper is as follows. Let us suppose: first, that we can predict the future course of the consumption of any particular commodity; second, that to produce this commodity we need a particular piece of equipment, or machine; third, that the survival characteristics for this type of machine are known; and, finally, that the age composition of the stock of this type of machine at the present time is also known. We may then ask the question: how many machines must we produce in each period in the future to maintain and increase the stock in such a way that at all times there are just enough machines to produce the amount of the commodity which we expect will be demanded? Evidently if we can solve this problem for one machine needed for one commodity we can in principle solve it for all machines needed to produce all commodities. By the use of current and capital input–output analysis we could further extend the solution to cover the machines needed to produce intermediate products and other machines. By this means we could work out the future course of demand for different types of machine and so obtain an aid to investment planning which would take account of the economy's present endowment of machines and future demand for their products.

In this paper we shall consider only the requirements for a single machine needed to produce a single commodity.

2. THE SIMPLEST CASE: FUTURE DEMAND CONSTANT

In this case it is assumed that we have the required number of machines for present needs and that all we have to do is to produce in each future year enough machines to maintain the stock at its present level.

Consider a set of survival rates, $s_0, s_1, \ldots$. The first of these, s_0, is the probability that a new machine is still alive on its first birthday; the second, s_1, is the probability that a machine that was alive on its first birthday is still alive on its second. These probabilities considered over the whole life-span of the machine are the elements of a vector $s = \{s_0\ s_1\ \ldots\}$. If the last element of s is $s_{\tau-1}$ then $s_{\tau-1} > 0$ and $s_\tau = 0$.

Whatever the age composition of the stock of machines, we shall keep this stock constant in number if we associate with the survival rates a set of birth or purchase rates $r_j = (1 - s_j)$ and purchase in each year an amount $\sum_j r_j n_j$ where n_j is the number of machines in the stock at the beginning of the year which are j years old.

This information, which all comes from the survival characteristics of the machines, can conveniently be arranged, following Leslie [97, 98], in a matrix, A say. If for simplicity we consider a machine which never passes its third birthday, then

$$A = \begin{bmatrix} r_0 & r_1 & r_2 \\ s_0 & 0 & 0 \\ 0 & s_1 & 0 \end{bmatrix} \tag{1}$$

where $r_j + s_j = 1$ and $r_2 = 1$. If $n_t = \{n_{0t} n_{1t} n_{2t}\}$ is a vector the element n_{jt} of which denotes the number of machines aged j at time t, then, assuming we wish to keep the number of machines constant, we see that

$$n_{t+1} = An_t \tag{2}$$

and, in general, that

$$n_{t+\theta} = A^\theta n_t \tag{3}$$

If we start with an arbitrary n_t we shall eventually approach an age composition which is characteristic of this type of machine.

To see what this age composition is, it is convenient to work with an imaginary type of machine which has certain fundamental characteristics similar to A but always lives through a complete life span. Suppose we multiply out all the survival rates and form a vector, h say, where, in the above example,

$$h = \{s_0 s_1 \quad s_1 \quad 1\} \tag{4}$$

The first element of h is the product of all the survival rates, the second is this divided by s_0 and the third is the second divided by s_1. Then we can define a matrix, A^* say, derived from A as follows

$$A^* = \begin{bmatrix} s_0 s_1 & 0 & 0 \\ 0 & s_1 & 0 \\ 0 & 0 & 1 \end{bmatrix} \begin{bmatrix} r_0 & r_1 & r_2 \\ s_0 & 0 & 0 \\ 0 & s_1 & 0 \end{bmatrix} \begin{bmatrix} 1/s_0 s_1 & 0 & 0 \\ 0 & 1/s_1 & 0 \\ 0 & 0 & 1 \end{bmatrix}$$

$$= \begin{bmatrix} r_0 & s_0 r_1 & s_0 s_1 r_2 \\ 1 & 0 & 0 \\ 0 & 1 & 0 \end{bmatrix} \tag{5}$$

or, in general,

$$A^* = \hat{h}A\hat{h}^{-1} \qquad (6)$$

where $\hat{h}$ denotes a diagonal matrix constructed from the vector h and $\hat{h}^{-1}$ denotes its inverse. Comparing A^* with A, we see that in A^* all the survival rates are 1 and that the purchase rates have been changed.

Since A and A^* are connected by a similarity transformation, they have the same characteristic equation. This means that

$$|\lambda I - A| = |\lambda I - A^*| = 0 \qquad (7)$$

where λ is a constant and I is the unit matrix, and so A and A^* have the same characteristic roots, λ_j say. Expanding the characteristic equation in powers of λ, we have, in this example,

$$\lambda^3 - r_0\lambda^2 - s_0 r_1 \lambda - s_0 s_1 r_2 = 0 \qquad (8)$$

Since there is only one change of sign in this equation, it follows by Descartes' rule of signs that there is only one characteristic root which is real and positive. This root is dominant and since

$$r_0 + s_0 r_1 + s_0 s_1 r_2 = 1 \qquad (9)$$

it is equal to 1. The corresponding characteristic vectors, which are arbitrary up to a multiplicative constant, may be written

$$i = \{1 \quad 1 \quad 1\} \qquad (10)$$

in the case of A^* and

$$\hat{h}^{-1}i = \{1/s_0 s_1 \quad 1/s_1 \quad 1\} \qquad (11)$$

in the case of A. Equations (10) and (11) give the stable age compositions corresponding to A^*-type and A-type machines. Once they are reached they will not be departed from. Thus, as we see from (11), it will eventually be necessary to buy in each period a proportion, π say, where

$$\pi = \{h_0(i'\hat{h}^{-1}i)\}^{-1} = 1/(1 + s_0 + s_0 s_1) \qquad (12)$$

of the stock in order to replace the machines that fail to survive. Here h_0 denotes the first element of h.

3. THE SECOND CASE: EXPONENTIAL GROWTH OF FUTURE DEMAND

In the previous case of constant future demand the dominant characteristic root of A was equal to 1. If we now turn to the case in which future demand is growing at a constant rate, we must consider the position when the dominant characteristic root is greater than 1. In place of A we will define a matrix, B say, obtained from A by multiplying the first row of A by a constant $\xi > 1$. Thus, in our example,

$$
B = \begin{bmatrix} \xi & 0 & 0 \\ 0 & 1 & 0 \\ 0 & 0 & 1 \end{bmatrix} \begin{bmatrix} r_0 & r_1 & r_2 \\ s_0 & 0 & 0 \\ 0 & s_1 & 0 \end{bmatrix}
$$

$$
= \begin{bmatrix} \xi r_0 & \xi r_1 & \xi r_2 \\ s_0 & 0 & 0 \\ 0 & s_1 & 0 \end{bmatrix} \tag{13}
$$

In a similar way we may define

$$
B^* = \hat{h} B \hat{h}^{-1} \tag{14}
$$

In place of (10) the dominant characteristic vector, l say, of B^* is

$$
l = \{\lambda_1^2 \quad \lambda_1 \quad 1\} \tag{15}
$$

and in place of (11) the dominant characteristic vector of B is

$$
\hat{h}^{-1} l = \{\lambda_1^2 / s_0 s_1 \quad \lambda_1 / s_1 \quad 1\} \tag{16}
$$

In these expressions λ_1 denotes the dominant characteristic root.

If the stock of machines grows steadily with an instantaneous growth rate ρ, say, then

$$
\lambda_1 = e^\rho \tag{17}
$$

and when a stable age composition has been reached

$$
\hat{h}^{-1} l_{t+1} = B \hat{h}^{-1} l_t = \lambda_1 \hat{h}^{-1} l_t \tag{18}
$$

Thus we can establish the connection between ξ and λ_1. For, in our example,

$$
\begin{bmatrix} \xi r_0 & \xi r_1 & \xi r_2 \\ s_0 & 0 & 0 \\ 0 & s_1 & 0 \end{bmatrix} \begin{bmatrix} \lambda_1^2 / s_0 s_1 \\ \lambda_1 / s_1 \\ 1 \end{bmatrix} = \begin{bmatrix} \lambda_1^3 / s_0 s_1 \\ \lambda_1^2 / s_1 \\ \lambda_1 \end{bmatrix} \tag{19}
$$

from the first line of which it follows that

$$
\xi = \frac{\lambda_1^3}{r_0 \lambda_1^2 + s_0 r_1 \lambda_1 + s_0 s_1 r_2} \tag{20}
$$

Thus if the stock of machines is to grow at a rate ρ, then the purchase rates appropriate to a stationary stock must be multiplied by ξ as given in (20). From (16), the proportion of the stock to be bought in each period, π, once a stable age distribution has been reached, is, in our example

$$
\pi = \frac{\lambda_1^2}{\lambda_1^2 + \lambda_1 s_0 + s_0 s_1} \tag{21}
$$

which reduces to the stationary case given in (12) if $\lambda_1 = 1$. Of this amount, π / ξ is required to maintain the stock and $(\xi - 1) \pi / \xi$ is needed to increase it.

4. THE THIRD CASE: SATURATION IN THE GROWTH OF FUTURE DEMAND

The cases considered in the two preceding sections have served the purpose of introducing the birth and death process in terms of which the required purchases in future years can be calculated. In practice these cases are of limited interest since it may reasonably be supposed that there is a maximum or saturation level to the demand for any commodity so that eventually any demand would cease to grow however much income might increase.

We shall see in section 6 below that in such a case the growth of demand over time may be approximated by a logistic curve. In this case the stock of machines required in production must also grow along a logistic curve. But this is precisely the pattern which will be followed by a population governed by the life-and-death process we have been considering provided that: (i) the initial age composition is appropriate to the stage of growth reached; (ii) the birth rates, r_j, are subject to a factor, η_t say, which varies geometrically with age; and (iii) the inverse of this factor is a linear function of the size of the population.

This means that in place of B we must now consider a matrix, C_t say, of the form

$$C_t = \begin{bmatrix} \xi\eta_t r_0 & \xi\eta_t^2 r_1 & \xi\eta_t^3 r_2 \\ s_0 & 0 & 0 \\ 0 & s_1 & 0 \end{bmatrix} \tag{22}$$

If $\eta_t = 1$, $C_t = B$ and so has the same characteristic roots and vectors as B. In the general case $\eta_t < 1$ and the elements of the characteristic vector of the transformed matrix, C_t^*, are powers not of λ_1 but of $\lambda_1\eta_t$. Thus the dominant characteristic vector of C_t, $\hat{h}^{-1}l_t$ say, is

$$\hat{h}^{-1}l_t = \{\lambda_1^2\eta_t^2/s_0 s_1 \quad \lambda_1\eta_t/s_1 \quad 1\} \tag{23}$$

If

$$\eta_t^{-1} = \alpha_0 + \alpha_1 v_t \tag{24}$$

where $v_t = i'n_t$ is the size of the population, or stock of machines, we are considering, then: (i) in a stationary state $\eta_t^{-1} = \lambda_1$; and (ii) as v_t tends to zero, η_t^{-1} must tend to 1. Thus we may write

$$\eta_t^{-1} = 1 + \frac{(\lambda_1 - 1)v_t}{\sigma} \tag{25}$$

where σ denotes the saturation level to which the population tends.

5. A SUMMARY OF THE TRANSITION MATRICES

The matrix A is a transition matrix which preserves the size of the population constant. Thus as we have seen

$$n_{t+1} = An_t \tag{26}$$

where $i'n_{t+1} = i'n_t$. If we wish the population to grow exponentially we must consider a matrix $B = XA$ where, in our example,

$$X = \begin{bmatrix} \xi & 0 & 0 \\ 0 & 1 & 0 \\ 0 & 0 & 1 \end{bmatrix} \tag{27}$$

If we put

$$n_{t+1} = Bn_t \tag{28}$$

then $i'n_{t+1} = \lambda_1 i'n_t = e^\rho i'n_t$. Finally if we wish the population to grow logistically we must consider a matrix $C_t = Y_t B = Y_t XA$ where, in our example,

$$Y_t = \begin{bmatrix} \eta_t^3 & \xi r_0 \eta_t(1-\eta_t^2)/s_0 & \xi r_1 \eta_t^2(1-\eta_t)/s_1 \\ 0 & 1 & 0 \\ 0 & 0 & 1 \end{bmatrix} \tag{29}$$

If we put

$$n_{t+1} = C_t n_t \tag{30}$$

or, in general,

$$n_{t+\theta} = C_{t+\theta-1} C_{t+\theta-2} \dots C_t n_t \tag{31}$$

where n_t has the characteristic age composition given by (23), then the successive values of $i'n_t, i'n_{t+1} \dots$ will follow a logistic curve.

6. TRANSITION MATRICES AND DEMAND ANALYSIS

In view of the importance of saturation, it is intuitively plausible that the Engel curve connecting the demand for a commodity with income should be sigmoid in character. In their classic study of budget data [1, 2], Aitchison and Brown showed that in fact all the commodity groups which they considered could be approximated by an Engel curve with the form of a lognormal integral. In a complementary study, Fisk [56] has shown that the same data can equally well be approximated by a logistic form of Engel curve. As is well known from works on biological assay, such as [54], these two forms of relationship are capable of giving approximately equivalent results. In the case of the lognormal integral, the assumption of a lognormal distribution of incomes with constant variance enables us to represent income by the average income, μ say. If we adopt a similar approximation in the case of the logistic Engel curve, we can proceed as follows. Let $v^* = v/\sigma$ denote consumption as a proportion of the saturation level of demand for the commodity investigated. Then with a logistic Engel curve connecting v^* and $\log_e \mu$, we have

$$v^*_\mu = \frac{1}{1 + \alpha \mu^{-\theta}} \tag{32}$$

If income is growing exponentially, then

$$\mu = \mu_0 e^{\rho t} \tag{33}$$

and so

$$v_t^* = \frac{1}{1 + \alpha^* e^{-\theta \rho t}} \tag{34}$$

where $\alpha^* = \alpha \mu_0^{-\theta}$. Thus demand is a logistic function of time.

According to the hypothesis we have been making, the stock of machines necessary to produce the consumption good must also grow logistically. For simplicity let us assume that we have measured consumption and machines in units such that the growth of the stock of machines is given by (34). Then we see that for the machines $\lambda_1 = e^{\theta \rho}$ and so the value of ξ can be obtained by inserting this value in (20). Also from (25) and (34) it follows that

$$\eta_t = \frac{1 + \alpha^* \lambda_1^{-t}}{\lambda_1 + \alpha^* \lambda_1^{-t}} \tag{35}$$

which can also be determined from the parameters of the Engel curve. Since C_t depends only on the survival characteristics of the machine, ξ and η_t, it follows that the population of machines, its age composition and so the numbers to be bought can be calculated, on the assumptions made, for each year in the future.

7. A RELAXATION OF THE ASSUMPTIONS

The foregoing theory is based on rather rigid assumptions relating in particular to the initial age composition of the stock of machines and the precise form of the growth of future demand. If either or both of these assumptions is to be relaxed, it is necessary to proceed step-by-step, using the A-matrix to calculate replacement requirements and the age composition of the machines in the initial stock at the end of the period, and using the forecasts of demand over this period to calculate the requirements for net investment. By these means the calculations can be made for any assumed age composition of the initial stock and any assumed growth pattern of future demand.

By similar step-by-step methods allowance can be made for the replacement after a certain date of machines which follow the original survival curve by new machines which follow a different one. For the new machines, a new A-matrix must be formed and enough of the new machines must be bought to replace the wastage of old and new machines and to make the appropriate increase in the stock. Evidently a difference in capacity between the old and new machines does not lead to any difficulty.

The question of when the old machines should be replaced by the new ones is outside the scope of this paper. Presumably this question would be settled by a comparison of discounted future costs but, however it is settled, the date of the changeover is taken as a datum in the method described above.

XIII

A DYNAMIC MODEL OF DEMAND

1. INTRODUCTION

In any empirical science, theory must lead to statements about the actual world which are not merely truisms, and these statements can be tested by comparing them with actual observations. For example, in its simplest form the theory of consumers' behaviour suggests that the amount demanded of any commodity will depend, given the tastes and habits of consumers, largely on their income and the structure of relative prices. Attempts to test this theory by the analysis of time series and of cross-section data as in [168] prove abundantly that it is a good starting point and, indeed, that many important conclusions about the behaviour of consumers can be established on this comparatively simple basis.

But the simplest models, though easy to work with, suffer from their very simplicity. It is our task as scientists to explain the phenomena we study and to do so in as simple terms as we can. But we must always be on our guard against oversimplification and so when we have got a model which provides a good approximation in many situations we must try to improve it so that it extends to a still wider range of situations; that is, it is more general. This process of going from simple models (first approximations) to more general ones (second and later approximations) is in fact the only way in which we can gain and extend useful knowledge. To refuse to consider simple models in a new subject because, after all, the world is a very complicated place, is sheer nihilism and if persisted in effectively brings any branch of science to a standstill.

The particular generalization which I shall consider in this paper is concerned with the adaptation of consumers to changing circumstances. The simple models of consumers' behaviour are static, which means that consumers are supposed to adjust themselves to changing circumstances instantaneously and do not become involved in processes of adjustment that take time. This is in many cases a reasonable assumption, particularly when it is remembered that our observations relate to years or quarters and not to days or hours. But in some cases we may believe that the period of adjustment is important and in these we should like a model which would enable us to take into account the rate at which consumers try to adjust themselves to changing curcumstances.

In the next section a dynamic model is given in terms of the demand for a

13

durable good. In the third section various generalizations of this model are given from which it is clear that it can easily be adapted to an analysis of the demand for perishable goods and indeed can provide a basis for the simultaneous analysis of the whole structure of consumption. Section 4 relates to various problems of calculation, and section 5 exemplifies the practical working of the model in terms of applications that have already been made. The paper ends with a brief concluding section.

2. A DYNAMIC MODEL OF THE DEMAND FOR A DURABLE GOOD

The following account relates to a dynamic model which Mr Rowe and I put forward in [164] though we had previously used a very similar model in studying the aggregate consumption function [163]. For the moment I shall concentrate on the economics of the model and leave questions of calculation and estimation until later.

Let q denote the quantity of some durable good which is bought in some period, say a year. These purchases all go either to maintaining the stock of the good existing at the beginning of the year or to increasing that stock. If we denote the use of the year (consumption) by u and the net addition to the stock by v, then

$$q \equiv u+v \tag{1}$$

The next equation provides a definition of consumption, u. It is assumed that the good lasts indefinitely but that each year it depreciates on account of use according to a reducing-balance depreciation formula. For example, suppose that the annual depreciation rate is 20 per cent. Then out of 100 new units the equivalent of 20 will be used up in the first year and the corresponding stock at the end of that year will be equivalent to 80 new units. The equivalent of 16 will be used up in the second year and the stock at the end of that year will be equivalent to 64 new units. In the third year, the use will be 12·8 units and the closing stock will be 51·2 units and so on indefinitely. Thus it is assumed that goods gradually wear out through use but are never actually scrapped. It is perfectly possible to allow for scrapping, as will be seen below, but this involves a knowledge of the survival characteristics of the good which is frequently not available.

If s denotes the stock at the beginning of the year, measured in the equivalent number of new units, and if $1/n$ is the rate of depreciation, then evidently the consumption of the year is at least as large as s/n. But in fact it is larger because some use is obtained from the purchases of the year, q. If these purchases were entirely concentrated at the beginning of the year, then the corresponding use would be q/n but in fact they are spread in some way through the year and so the use is equal to q/m where $m \geqslant n$. Thus

$$u \equiv \frac{s}{n}+\frac{q}{m} \tag{2}$$

If it is assumed that purchases are spread evenly through the year then we shall see that m is related to n in quite a simple way.

These relationships are truisms in the sense that we choose to define our variables so that they hold in all circumstances. The next relationship is not a truism but is a hypothesis about consumers' behaviour. It is based on the concept of an equilibrium stock, s^* say, which if attained in any given situation would leave consumers content, in the sense that they would wish neither to increase nor to diminish it. If s^* is different from s it is assumed that consumers will take steps to bring the two variables into equality, increasing their stock if $s^* > s$ and diminishing it if $s^* < s$. It is assumed that in general they cannot close the gap immediately and that what they actually do is to close a proportion of it, r say, in a period. Thus

$$v = r(s^* - s) \tag{3}$$

Now s^* is not directly observable whereas all the other variables are either observable or can be calculated from the definitions. It is assumed therefore that s^* depends on income and the price structure: for example, we might have

$$s^* = \alpha + \beta\left(\frac{\mu}{\pi}\right) + \gamma\left(\frac{p}{\pi}\right) \tag{4}$$

where μ is money income, p is the price of the good and π is an index of all other goods and services bought by consumers. Unless the good (or group of goods) being analysed is important in the consumers' total budget, π will usually be indistinguishable from an index of retail prices.

By combining (1), (2) and (3) we can express purchases and consumption in terms of equilibrium stocks and actual stocks. Thus

$$q = k[rns^* + (1 - rn)s] \tag{5}$$

where $k \equiv m/n(m-1)$; and

$$u = k[r^* s^* + (1 - r^*)s] \tag{6}$$

where $r^* \equiv rn/m$. Thus both purchases and consumption are proportional to certain weighted averages of actual and equilibrium stocks. These variables, q and u, can be expressed in terms of observable variables by substituting for s^* from (4) into (5) and (6).

We can now see that this model is indeed a generalization of the static model. For if $rn = 1$ the second term on the right-hand side of (5) disappears and q is proportional to s^*, that is to say it depends in a static way on income and the price structure.

But there are now two additional possibilities. First, $rn < 1$; in this case the opening stock, s, exerts a positive influence on q. Second, $rn > 1$; in this case s exerts a negative influence on q. The interpretation of these cases will be clear when we have examined the expressions for long and short term responses.

The ultimate level of purchases, q^* say, associated with a given level of s^* is equal to the level of consumption, $u^* \equiv ks^*$ say, which is associated with s^* since, when $s = s^*$, $v = 0$. So the long-term partial derivative of purchases with respect to income is

$$\frac{\partial q^*}{\partial \mu} = \frac{k\beta}{\pi} \tag{7}$$

while the corresponding short-term derivative can be seen from (5) to be

$$\frac{\partial q}{\partial \mu} = \frac{rnk\beta}{\pi} \tag{8}$$

So the ratio of the short to the long-term response is

$$\frac{\partial q}{\partial q^*} = rn \tag{9}$$

Thus if $rn > 1$ the immediate effect of a change in s^* is greater than the long-term effect, and conversely if $rn < 1$. So if the ratio of the rate of adjustment to the rate of depreciation is greater than one, a change in s^* will have a larger effect immediately than in the long run. The reason is that, given n, consumers are in this case trying to adjust quickly and so add rapidly to their stocks with the consequence that they buy more in the short run than will be ultimately needed merely to maintain their new stocks. In the converse case they will adjust slowly, starting by buying very little and ultimately increasing their purchases to the level needed to maintain their new level of stocks.

Clearly there is no difficulty in restating these results in terms of elasticities.

It is frequently convenient in demand analysis to express the static equation in a log-linear rather than a simple linear form. In this case (4) takes the form

$$s^* = \alpha \left(\frac{\mu}{\pi}\right)^\beta \left(\frac{p}{\pi}\right)^\gamma \tag{10}$$

in which case it is convenient to change (3) to

$$v = \frac{ms}{n}\left[\left(\frac{s^*}{s}\right)^{r^*} - 1\right] \tag{11}$$

from which it follows that

$$u = ks^{*r^*} s^{1-r^*} \tag{12}$$

and

$$q = kms^{*r^*} s^{1-r^*} - \frac{ms}{n} \tag{13}$$

By comparing (6) and (12) we see that a geometric average has replaced an arithmetic average, and comparing (5) with (13) we see that the expression for q is a little more complicated than in the linear case.

3. GENERALIZATIONS

A number of generalizations follow easily from what has been said in the last section; let us look at some of these.

(a) *The case of perishable goods.* These goods correspond to the limiting case in which $n \to 1$. It is shown in [164] that

$$\lim_{n \to 1} ks = E^{-1}q \tag{14}$$

where E is an operator such that $E^{-\theta}q_t \equiv q_{t-\theta}$. Thus, for example, ks in (5) and (6) is replaced by $E^{-1}q$, that is by the purchases of the preceding period.

(b) *Time lags in the determining variables.* In (4) it is assumed that s^* is determined by the current value of income and the price structure. In practice consumers may adjust in terms of some average of past incomes or even of some kind of income projection. Such considerations may easily be incorporated in the model. Thus let $\mu/\pi \equiv \rho$. Then in place of ρ in (4) we might substitute

$$\rho^* \equiv \eta\rho + (1-\eta)E^{-1}\rho \tag{15}$$

If we used (3) as an estimating equation the regression estimators would relate to $r\alpha$, $r\beta\eta$, $r\beta(1-\eta)$, $r\gamma$ and r, from which we could estimate $\beta\eta$ and $\beta(1-\eta)$ and therefore β and so η.

(c) *A variable rate of adjustment.* Again assume, as will normally be the case, that (3) is used as an estimating equation. Since r is a factor of the right-hand side of (3), it follows that, if x is an observable variable, we can make r a function of x by replacing s^* and s by xs^* and xs. Two examples of this useful device may be given.

First, suppose we wish to allow for hire-purchase conditions, that is to say the possibility of paying only a proportion, ξ say, of the purchase price at the time of purchase plus a number of regular instalments spread over τ time periods. Then we might define hire-purchase conditions x, as

$$x \equiv \xi + \frac{1-\xi}{\tau} \tag{16}$$

If we wished to allow x to influence the equilibrium stock we could do this by dividing p/π in (4) into two parts: xp/π and $(1-x)p/\pi$ and treating these as separate variables. If we wished to allow x to influence the rate of adjustment we could replace r by r^{**}/x where r^{**} is a constant. From (3) it follows that in this case

$$xv = r^{**}(s^* - s) \tag{17}$$

provides a satisfactory estimating equation. In practice, to keep (3) and (17) comparable, it is convenient to take $xv/\bar{x}$, where $\bar{x}$ is the mean value of x, as the dependent variable.

Second, we may have to analyse the demand for a comparatively new commodity which consumers are only gradually learning to appreciate. In this case consumers are finding out the potentialities of the good and only a limited number, the pioneers, are seriously in the market. Accordingly the average rate of adjustment will be low because even if the pioneers are adjusting in the way that the whole community will eventually come to adopt, they are, initially, only a small, though growing, part of the whole consuming public. In this case we might wish to make the rate of adjustment dependent on the proportion of the population which already possessed the new good.

(*d*) *Allowance for scrapping.* In section 2 it was assumed that durable goods last for ever but are subject to a continuous depreciation through use so that the contribution of an individual unit to the total stock follows, through time, an exponential decay curve. This is, in a sense, a limiting assumption. We might equally well have assumed that a good does not depreciate at all through use; that is an old good is just as satisfactory as a new one except for the fact that it is more likely to fall to pieces and so be unusable. The reason why this alternative assumption was not made in section 2 is that to make use of it we need to know the specific mortality rate of goods in each year of their life. Typically we do not have this information but if we had it we could proceed as follows.

Let λ denote the mortality rate in the period in which the good is bought and let l denote a vector of mortality rates for goods bought last year, the year before that and so on. Then we might define consumption as

$$u \equiv l's + \lambda q \tag{18}$$

where s denotes a vector of stocks of different ages existing at the beginning of the period. If (2) is replaced by (18) we have what may be called a replacement model rather than a depreciation model. But now we need not two parameters, m and n, but a larger number, λ and l, depending on the survival characteristics of the good. Typically, survival rates will follow, approximately, a logistic decay curve; that is to say in the early years of life most units will survive, in middle life there will be a sharp decline in survival rates from year to year, and finally survival rates will tend to zero.

In fact we can combine the depreciation and replacement models so as to give a better approximation to reality. In this case we define consumption as follows. In the first place it is equal to the amount scrapped out of current purchases plus the depreciation of the period on the amount bought in the period which survives. This component is $[\lambda + (1-\lambda)/m]q$. From a written-down stock at the beginning of the period, which we might denote by $\hat{w}s$ (w being a vector which embodies any depreciation formula), the total mortality in the year is $l'\hat{w}s$. The depreciation of the year on units in the written-down value of the stock at the beginning of the year is $d'(I - \hat{l})\hat{w}s$, where I denotes the unit matrix. So this component is $l'\hat{w}s + d'(I - \hat{l})\hat{w}s$. Thus for this model

$$u \equiv [l' + d'(I - \hat{l})]\hat{w}s + [\lambda + \delta(1 - \lambda)]q \tag{19}$$

where $\delta \equiv 1/m$. Consideration will show that (19) generalizes the models based on (2) and (18).

(*e*) *A system of demand relationships.* We have so far considered a single demand equation but for some purposes we may wish to consider a consistent system of demand relationships. A means of doing this is set out in [148]. This system can be made dynamic by allowing the basic equation to determine equilibrium consumption rather than actual consumption. The characteristics of this system, which will not be described here, can be found in [162].

(*f*) *The aggregate consumption function.* The relationship of consumers' total expenditure to their income is important for many purposes of economic policy. An attempt to investigate it with the help of an early version of the system just mentioned is given in [163].

(*g*) *A generalization of the input–output model of Leontief.* The dynamic system just referred to generalizes the input–output model of Leontief since, in principle, it could be applied to every sector of a Leontief input–output matrix. If this were done every sector would respond to the price structure as well as to its revenue. Some thoughts on such a model are contained in [149]. But the estimation procedure used there is inefficient and for the proper procedure reference should be made to [25].

4. CALCULATIONS

The model described in this paper gives rise to numerous problems of calculation and estimation which must be faced if it is to be applied. Let us look at some of these.

(*a*) *The calculation of stocks from past purchases.* If we look at equations (1) and (2) we see that there are six symbols: q, u, v, s, n and m. Let us assume that n is given; we shall see shortly that this enables us to calculate m. In addition we may assume that q is given. Thus we have three unknowns, u, v and s, and only two equations. So we need a third equation before the unknowns can be calculated.

This equation is an identity connecting stocks and flows, namely

$$Es \equiv s + v \tag{20}$$

that is: stocks at the end of the year equal stocks at the beginning plus net investment in the year. From (1), (2) and (20)

$$Es \equiv s+q-u$$

$$\equiv \left[\frac{n-1}{n}\right]s + \left[\frac{m-1}{m}\right]q$$

$$\equiv as+bq \tag{21}$$

say, where $a \equiv (n-1)/n$ and $b \equiv (m-1)/m$. If (21) is multiplied by the operator E^{-1} and if a continuous substitution is made for $E^{-\theta}s, \theta = 1, 2, \ldots,$ there results

$$s \equiv \frac{b}{a} \sum_{\theta=1}^{\infty} a^{\theta} E^{-\theta} q$$

$$\equiv \frac{b}{a} \sum_{\theta=1}^{\tau} a^{\theta} E^{-\theta} q + a^{\tau} E^{-\tau} s \tag{22}$$

if the calculation has to be started with an assumed stock, $E^{-\tau}s$, available τ time periods ago.

(*b*) *The relationship of m to n.* If it is assumed that purchases are spread evenly through each period, then the relationship of m to n is given in [164]. In a slightly different form, this may be written as

$$m = \sum_{\theta=1}^{\infty} (1/\theta n^{\theta}) / \sum_{\theta=2}^{\infty} (1/\theta n^{\theta}) \tag{23}$$

This expression converges rapidly to the value $2n - \frac{1}{3}$ as n increases.

(*c*) *The time taken for a good to depreciate through use to 10 per cent of its original amount.* Let n^* denote the number of time periods that must elapse before a good with depreciation rate $1/n$ is 90 per cent depreciated. Then, as shown in [164],

$$n^* = 2 \cdot 3 / \sum_{\theta=1}^{\infty} (1/\theta n^{\theta})$$

$$= 2 \cdot 3n - 1 \cdot 2 \tag{24}$$

approximately. Thus for example if $n = 5$, $n^* = 10 \cdot 3$ approximately.

(*d*) *Internal estimates of* $1/n$. It has been shown by Nerlove in [115] that the depreciation rate can be estimated from the observations. The procedure is as follows. By combining (1), (2) and (20) we see that

$$q = \frac{m-1}{m} \left\{ Es - \left[\frac{n-1}{n}\right]s \right\}$$

$$= k[1 + (n-1)\Delta] Es \tag{25}$$

where $\varDelta \equiv 1 - E^{-1}$. Thus

$$\varDelta q = k[1+(n-1)\varDelta]v$$
$$= k\Omega v \qquad (26)$$

say. Thus, by applying the transformation Ω to the variables on which s^* depends, we obtain a suitable estimating equation. By trying different values of n we can then select that one which gives the highest correlation in the regression equation. Some results will be found in [166].

(e) *The combination of data with different time units.* We may wish to combine data with different time units; for example in Britain we have quarterly data for the post-war period but only annual data for the pre-war period. If we want to combine these observations we have to solve an aggregation problem. This is done for a linear model in [164]. This solution provides a linear approximation for non-linear models.

5. APPLICATIONS

It may be of interest to indicate some of the results obtained from the model discussed in this paper.

(a) *Analyses for individual groups of durable goods.* In [164] analyses were given for clothing and for household durable goods. These were carried out for the inter-war period and for it and the post-war years to 1955 combined. A log-linear model was used and variables were transformed to first differences. The calculations were carried out with several different values of the depreciation rate. The highest correlations were obtained with $n=1$ for clothing and $n=4$ for household durable goods. When short- and long-term elasticities were calculated it was found that the long-term income elasticities were very similar to those obtained from household budgets collected in 1937–39. In other respects the basic features of demand responses appeared quite similar in the inter-war and post-war periods.

More recently a further analysis has been made of the two major categories of household durable goods: furniture and floor coverings, and radio and electrical goods. In each case it was assumed that $n=4$. Quarterly data for the years 1953 through 1958 were used and the model was a linear one exactly as set out in this paper. The main purpose of these new analyses was to test the importance of hire-purchase in the ways suggested in the earlier part of this paper. The main results can be summarized in the following table.

From this table we see that income is a very important influence both in the short and long run whereas price has very little influence for the group as a whole. We also see that by introducing hire-purchase we improve the analysis but that having introduced it in one way or the other we cannot easily decide which formulation is the better. The main effect of introducing

DEMAND ELASTICITIES FOR FURNITURE AND FLOOR COVERINGS

UNITED KINGDOM, 1953–58

	Income elasticities		Price elasticities		Proportion of the variance of purchases accounted for
	Short-term	Long-term	Short-term	Long-term	
No allowance for hire-purchase	4·70	1·90	0·05	0·02	0·86
Hire-purchase affecting equilibrium stock	3·01	2·13	−0·26	−0·18	0·92
Hire-purchase affecting the rate of adjustment	3·51	2·03	0·14	0·08	0·93
Hire-purchase affecting both	2·82	2·11	−0·08	−0·06	0·94

hire-purchase in either way is to reduce the high value of the short-term income elasticity which appears if it is left out. It is interesting to note that according to the budget investigations of 1937–39 the income elasticity for this group was about 2·4.

(*b*) *Comparison of pre-war and post-war demand structures.* In [165] a number of analyses are given for both the inter-war and post-war periods. It is found that in many cases the structural parameters are quite close. It appears to be generally true that the rate of adjustment is relatively low (of the order of 0·3 in annual terms) for durable goods and relatively high (of the order of 1 in annual terms) for perishable goods. In these analyses a successful attempt was made to trace the influence of temperature and rainfall on the demand for certain goods.

(*c*) *Dynamic systems of demand relationships.* In [162] an account is given of the first attempt to work with a complete dynamic system of relationships. The results are encouraging and in fact give quite a good account of observed variations in the seven commodity groups among which personal consumption is subdivided. At the time when this study was made it was not possible to handle more than seven commodity groups and the two-stage iterative computing procedure was not in a technically efficient form. These difficulties have now been overcome to some extent: we can now handle up to thirteen groups and the computer programme has been radically improved.

The analysis in [162] gives for the first time a complete matrix of elasticities of substitution which is symmetric (that is satisfies the Slutsky condition). It shows that with the relatively large groups used the sub-stitution effects of changes in the price structure are relatively small. It also suggests that a large part of consumers' expenditure is, as one might expect, in a sense committed in advance. This in turn suggests that large, sudden changes in consumption patterns are not very likely to occur on a broad front, and this is in accordance with historical observations.

An analysis of thirteen commodity groups has just been completed with the help of the new programme. The results are interesting but have not yet been fully analysed.

6. SUMMARY AND CONCLUSIONS

I have tried to summarize some recent work in dynamic demand analysis which is giving good results but which is still at a comparatively early stage of development. I have concentrated on a formulation of the basic model which emphasizes the economic assumptions involved and have indicated some generalizations which readily suggest themselves and which illustrate the potentialities of the model. I have described some calculations which are useful and may not be immediately obvious, but I have not gone in any detail into statistical and computing techniques. These are all well known so far as the simple models are concerned and extremely difficult and by no means fully solved for the complete system of demand equations. Finally I have outlined the main results which have so far been obtained.

XIV

THE CHANGING PATTERN OF CONSUMPTION

1. INTRODUCTION

The object of this paper is to describe the system of demand functions which we are using in our model of British economic growth [30]. The purpose of these functions is to enable us to divide a given total of consumption into its constituent commodities and so to take us a step nearer our goal, a balanced statement of the economy in a future year.

Our method has been to start with a simple system of relationships which possesses a number of generally accepted theoretical properties and then to elaborate this model to take account of what we know about consumption patterns in the past. The development of the model is thus an example of the iterative process of induction and deduction commonly found in scientific work.

In the following section I shall explain the economics of the model and in section 3 I shall show how we have solved or how we propose to solve the various statistical problems to which the model gives rise. In section 4 I shall set out some of the results and in section 5 I shall give a brief summary of conclusions.

2. THE ECONOMICS OF THE SYSTEM

The system of demand functions I am discussing relates to the average consumer. In its simplest form, it can be described as follows.

The average consumer has a concept of the standard of living he expects to be able to achieve. This concept is expressed in terms of a set of quantities whose elements are the amount of each commodity which must be consumed if the standard of living is to be realized. The average consumer buys these quantities notionally at their current market prices and then compares the total cost of what he has bought with the amount of money he allows himself for spending on consumption. If he finds he has some money over he allocates this to the different commodities in certain fixed proportions; if he finds that he has overspent the money available he reduces his expenditure on the different commodities by applying the same proportions to the amount of overspending.

Let us stop at this point and express this set of relationships in algebra.

Let p denote the vector of commodity prices and e the vector of quantities bought by the average consumer; then $\hat{p}e$ denotes the vector of expenditures, a circumflex accent on a vector being used to denote a diagonal matrix formed from it. Let μ denote total expenditure, the elements of a vector c denote the constituents of the basic standard of living and the elements of b denote the proportions in which uncommitted expenditure is devoted to the different commodities. Then

$$\hat{p}e = \hat{p}c + b(\mu - p'c)$$
$$= b\mu + (I - bi')\hat{c}p \tag{1}$$

where i and I denote respectively the unit vector and the unit matrix and a prime denotes transposition. Since $i'b \equiv 1$, premultiplication of (1) by i' yields the identity $p'e \equiv \mu$. Premultiplication of (1) by $\hat{p}^{-1}$, the inverse of $\hat{p}$, shows that the elements of e are homogeneous linear functions of degree zero in μ and p, so that e is unchanged if μ and p are changed to $\lambda\mu$ and λp where λ is any positive constant. The equation for an element, β say, of e can be written in the form

$$e_\beta = (1 - b_\beta)c_\beta + b_\beta(\mu - \sum_{\gamma \neq \beta} p_\gamma c_\gamma)/p_\beta \tag{2}$$

from which it follows that

$$\frac{\partial e_\beta}{\partial \mu} \cdot \frac{\mu}{e_\beta} = \frac{b_\beta \mu}{p_\beta e_\beta} \tag{3}$$

$$\frac{\partial e_\beta}{\partial p_\beta} \cdot \frac{p_\beta}{e_\beta} = -\frac{b_\beta\left(\mu - \sum_{\gamma \neq \beta} p_\gamma c_\gamma\right)}{p_\beta e_\beta} \tag{4}$$

and

$$\frac{\partial e_\beta}{\partial p_\gamma} \cdot \frac{p_\gamma}{e_\beta} = -\frac{b_\beta p_\gamma c_\gamma}{p_\beta e_\beta} \tag{5}$$

so that

$$\frac{\partial e_\beta}{\partial \mu} \cdot \frac{\mu}{e_\beta} = -\sum_{\gamma=1}^{v} \frac{\partial e_\beta}{\partial e_\gamma} \cdot \frac{p_\gamma}{e_\beta} \tag{6}$$

The price elasticities in (4) and (5) are elasticities along uncompensated demand curves and can be divided into an income effect and a substitution effect. Thus if we denote by w_γ the proportion of total expenditure devoted to commodity γ, so that

$$w_\gamma \equiv \frac{p_\gamma e_\gamma}{\mu} \tag{7}$$

and by $s_{\beta\gamma}$ the elasticity of substitution between commodities β and γ, then

$$s_{\beta\gamma} = -\frac{(\delta_{\beta\gamma} - b_\beta)b_\gamma(\mu - p'c)}{w_\beta w_\gamma \mu} \tag{8}$$

where $\delta_{\beta\gamma}=1$ if $\beta=\gamma$ and is otherwise equal to zero. Thus we can write

$$\frac{\partial e_\beta}{\partial p_\gamma}\cdot\frac{p_\gamma}{e_\beta} = w_\gamma\left[s_{\beta\gamma}-\left(\frac{\partial e_\beta}{\partial \mu}\cdot\frac{\mu}{e_\beta}\right)\right] \tag{9}$$

By appropriate substitutions (9) can be reduced to (4) if $\beta=\gamma$ and to (5) if $\beta\neq\gamma$. The first term, w_γ, $s_{\beta\gamma}$, on the right-hand side of (9) measures the substitution effect of a change in the price of γ on the demand for β and the second term measures the income effect. This is Slutsky's equation, and, as can be seen from (8), Slutsky's condition that $s_{\beta\gamma}=s_{\gamma\beta}$ is satisfied by this system of equations.

In theory the own-elasticities of substitution, $s_{\beta\beta}$, must be negative. If, as seems reasonable, we assume that uncommitted expenditure, $\mu-p'c$, is positive, this condition requires that $0<b_\beta<1$. Thus inferior goods are ruled out. But with this restriction on the elements of b it follows that $s_{\beta\gamma}>0$ for all $\beta\neq\gamma$. Thus complementary goods are ruled out too, and the system can only represent a set of commodities, or commodity groups, that are substitutes for one another. With a careful choice of commodity groups, this may not be a serious limitation in practice.

There are, however, other limitations to this simple formulation which are serious. The most obvious, which I shall discuss in some detail in this paper, is that so far the elements of b and c have been assumed constant. As time progresses, the average consumer's conception of his standard of living is likely to change; and with it his allocation of uncommitted expenditure is likely to change too. What can we do about this?

Fortunately, it is not very difficult to allow for systematic changes in b and c. The formal properties of the model are not affected if the parameters are made functions of predetermined variables. The simplest possibility is to make them linear functions of time. Thus, at time θ, we should have

$$b_\theta = b^*+\theta b^{**} \tag{10}$$

and

$$c_\theta = c^*+\theta c^{**} \tag{11}$$

say.

The introduction of these linear trends removes the main rigidity of the original formulation in (1). Nevertheless, cases arise in which linear trends are too crude an approximation. For example, one of the categories we have used in our empirical work is transport which includes expenditure on cars and their running expenses as well as on public transport. At the beginning of the century, cars were of negligible importance; by the nineteen-thirties they were generally accepted but were still too expensive for many people; nowadays they are coming to be bought by the whole community. As a consequence the trends in the parameters for transport have tended to move along an accelerating curve. Such a tendency can be represented in the model by adding quadratic terms to (10) and (11).

We shall see in section 4 below that we can improve by this means the ability of the model to describe past observations. But in doing this we risk the possibility that our projections will be less reliable than they would have been with linear trends. For example, the accelerating trends for transport will eventually pass through a point of inflection and begin to slow down. In other cases we may find that a quadratic trend passes through its maximum or minimum near the end of the period of observation and so will change its direction in the period of projection. Though possible in exceptional cases, for example the gradual disappearance of the carriage after the advent of the motor car, such changes of direction are, in general, not very plausible. They can be avoided by giving up time as the variable on which the parameters depend and by making them functions of the past history of the branch of demand to which they relate. For example if e_θ^* denotes a vector of three-year moving averages of the components of consumption ending in year $\theta - 1$, then we could replace (10) and (11) by

$$b_\theta = b^* + i' e_\theta^* b^{**} \tag{12}$$

and

$$c_\theta = c^* + e_\theta^* c^{**} \tag{13}$$

We shall see in the next section that this formulation presents no more statistical difficulties than the earlier ones.

Having thus freed the basic model from its failure to allow for changing tastes and habits, we must now consider another limitation. Even if the parameters change systematically through time, the model still implies that consumers are capable of rapid adaptation so that in each year they are in equilibrium. This is probably a reasonable assumption in the case of perishable goods but it is certainly not reasonable in the case of major durable goods which involve a large initial expenditure. As explained in [30] the method described in [164, XIII] can be adapted to the present model. A computing sequence for this extension of the model is given in [160]. I shall not discuss this problem further here as we have not so far made use in our calculations of this extension of the model.

3. THE STATISTICS OF THE SYSTEM

In applying the system of equations just described we have used annual observations over the period 1900 to 1960. The first results, using linear trends in the parameters, were set out in [160]. In this paper only broad groups were analysed and no attempt was made to base the estimates of the parameters on cross-section data as well as time series. Accordingly, in this section I shall consider three problems: (*a*) a computing sequence for the model consisting of (1), (12) and (13); (*b*) a decomposition of the model to enable the subgroups of main groups to be analysed; and (*c*) the combination of cross-section data and time series in estimating the parameters.

(a) *The computing sequence.* In order to estimate the parameters b^*, b^{**}, c^* and c^{**} we have used an iterative two-stage least-squares procedure. If we consider the model consisting of (1), (12) and (13), the computing sequence is as follows.

We begin by guessing values of b^* and b^{**} which I shall denote by b_0^* and b_0^{**}. The values of the elements of b_0^* are the average expenditure proportions; those of b_0^{**} are zero.

We then form a vector of type $\nu \times 1$, y_θ say, as follows

$$y_\theta \equiv \hat{p}_\theta e_\theta - (b_0^* + i' e_\theta^* b_0^{**}) \mu_\theta \tag{14}$$

and a matrix of order ν, Y_θ say, as follows

$$Y_\theta = [I - (b_0^* + i' e_\theta^* b_0^{**}) i'] \hat{p}_\theta \tag{15}$$

Apart from a random element, y_θ and Y_θ are connected by the relationship

$$y_\theta = [Y_\theta : Y_\theta \hat{e}_\theta^*] \begin{bmatrix} c^* \\ \cdots \\ c^{**} \end{bmatrix} \tag{16}$$

If we now define

$$y \equiv \{y_1, y_2, \ldots, y_\tau\} \tag{17}$$

$$Y \equiv \{Y_1, Y_2, \ldots, Y_\tau\} \tag{18}$$

and

$$Y^* \equiv \{Y_1 \hat{e}_1^*, Y_2 \hat{e}_2^*, \ldots, Y_\tau \hat{e}_\tau^*\} \tag{19}$$

we can write, apart from a random element,

$$y = Xg \tag{20}$$

where $X \equiv [Y : Y^*]$ and $g \equiv \{c^* : c^{**}\}$.

The least squares estimator, g_1, of g is

$$g_1 = (X'X)^{-1} X'y \tag{21}$$

Given g_1 we can form a vector of type $\nu \times 1$, w_θ say, as follows

$$w_\theta \equiv \hat{p}_\theta [e_\theta - (c_1^* + \hat{e}_\theta^* c_1^{**})] \tag{22}$$

and a matrix of order ν, W_θ say, as follows

$$W_\theta \equiv [\mu_\theta - \hat{p}_\theta'(c_1^* + \hat{e}_\theta^* c_1^{**})] I \tag{23}$$

It will be noticed that W_θ is a scalar matrix. Apart from a random element, w_θ and W_θ are connected by the relationship

$$w_\theta = [W_\theta : W_\theta i' e_\theta^*] \begin{bmatrix} b^* \\ \cdots \\ b^{**} \end{bmatrix} \tag{24}$$

so that if we define

$$w \equiv \{w_1, w_2, \ldots, w_\tau\} \tag{25}$$

$$W \equiv \{W_1, W_2, \ldots, W_\tau\} \tag{26}$$

and

$$W^* \equiv \{W_1 i' e_1^*, W_2 i' e_2^*, \ldots, W_\tau i' e_\tau^*\} \tag{27}$$

we can write, apart from a random element,

$$w = Zh \tag{28}$$

where $Z \equiv [W : W^*]$ and $h \equiv \{b^* : b^{**}\}$. The least squares estimator, h_1, of h is

$$h_1 = (Z'Z)^{-1} Z' w \tag{29}$$

Given h_1 we can return to (16), replace b_0^* and b_0^{**} by b_1^* and b_1^{**} and calculate the next approximation $g_2 \equiv \{c_2^* : c_2^{**}\}$ of g. We continue in this way until the process converges.

In estimating h, the system breaks down into ν separate equations since W_θ and W_θ^* are diagonal matrices. At the same time the adding-up theorem ensures that $i' b^* = 1$ and $i' b^{**} = 0$ for the estimated values of b^* and b^{**}. In estimating g on the other hand, since Y_θ and Y_θ^* are not diagonal matrices, the ν equations all contribute to a single, average estimator of g.

(b) The analysis of subgroups. A feature of the system I am describing is that it is decomposable: once we have analysed total consumption divided into a certain number of main groups, we can then carry out exactly similar analyses on the components of each of the main groups. If necessary we can continue this process in a hierarchy of sub-analyses.

The method is as follows. The equations for group j in a complete system can be written as

$$\hat{p}_j e_j = \hat{p}_j c_j + b_j(\mu - p' c) \tag{30}$$

where the suffix j denotes that the vector to which it is attached contains elements relating only to group j. If we premultiply (30) by i' we obtain

$$p_j' e_j = \mu_j$$
$$= p_j' c_j + i' b_j(\mu - p' c) \tag{31}$$

so that

$$\mu - p' c = (i' b_j)^{-1} (\mu_j - p_j' c_j) \tag{32}$$

If we substitute for $\mu - p' c$ from (32) into (30) we obtain

$$\hat{p}_j e_j = \hat{p}_j c_j + b_j (i' b_j)^{-1} (\mu_j - p_j' c_j) \tag{33}$$

From an analysis of the main groups, we can obtain estimates of $i' b_j$ and $i' c_j \equiv {}_j c$, say. From the analysis of the components of group j we can obtain estimates of c_j and $b_j(i' b_j)^{-1}$. Thus the whole system will fit together consistently provided that $i' c_j$ in the sub-analysis is equal to ${}_j c$ in the main analysis. To ensure this equality we must carry out stage 1 of the sub-analysis subject to this constraint.

14

If, for example, we consider the version of the system in which the parameters are linear functions of time, then we must ensure that $_jc^* = i'c_j^*$ and that $_jc^{**} = i'c_j^{**}$. If λ_1 and λ_2 denote two undetermined multipliers, then the normal equations of stage 1 of the computing sequence for the sub-analysis take the form

$$
\begin{bmatrix}
Y_j'Y_j & \theta Y_j'Y_j & \vdots & i & 0 \\
\theta Y_j'Y_j & \theta^2 Y_j'Y_j & \vdots & 0 & i \\
\cdots\cdots\cdots\cdots\cdots & & \vdots & \cdots & \\
i' & 0 & \vdots & 0 & 0 \\
0 & i' & \vdots & 0 & 0
\end{bmatrix}
\begin{bmatrix}
c_j^* \\
c_j^{**} \\
\cdots \\
\lambda_1 \\
\lambda_2
\end{bmatrix}
=
\begin{bmatrix}
Y_j'y_j \\
\theta Y_j'y_j \\
\cdots \\
_jc^* \\
_jc^{**}
\end{bmatrix}
\tag{34}
$$

The equations in the second stage of the computing sequence are unaffected.

(c) Cross-section data and time series. Although our empirical work on the model I am describing has been based so far on time series, it would obviously be desirable to check the conclusions derived from it with the estimates of the relationships between individual expenditures and total expenditure derivable from budget studies. Since there are obvious difficulties in comparing derivatives or elasticities obtained from time series with those obtained from budgets, the first step is to make independent calculations on the two bases and find out if they differ significantly. If they do, we must conclude that apparently comparable measures are in fact not really comparable. A possible reason for this which seems to be borne out by a limited amount of analysis [165] is that for some types of good and, in particular, for durable goods, long-term total expenditure elasticities are typically different from short-term elasticities. Analyses of time-series which do not concern themselves with the time needed for adaptation, may reasonably be supposed to yield estimates of short-term elasticities. Elasticities from budgets on the other hand may better approximate to long-term elasticities. We could test this approximation by using the dynamic version of the model based on time series. If we find that the two sets of estimates are, on the whole, not very different we should combine them to give better estimates of the parameters. Following Durbin [52], there are two ways of doing this.

(i) We estimate a time-series of b from the budget studies, supposing these to be sufficiently numerous, and use these as extraneous estimators in (1). With this information we can rewrite (1) as

$$
u = Vc
\tag{35}
$$

where $u \equiv \hat{p}e - b\mu$ and $V = (I - bi')\hat{p}$. Thus we could estimate c from the equation

$$
c = (V'V)^{-1}V'u
\tag{36}
$$

In all likelihood the budget studies of the past will not be sufficiently numerous for this purpose. If this is so we ought to carry out the fitting simultaneously, subject to an appropriate constraint on the time-form of b.

(ii) Statistically speaking, the method just suggested does not make an efficient use of the available data. A method which made the fullest use of the data from both sources would have to follow the lines suggested in section 3 of [52].

By these means it should be possible to build up a fairly detailed picture of the structure of demand consistent with all the information about the past that we possess. In the course of doing this we may be led away from the particular simple approximations that I have described. Our problems, then, are: first to find a suitable specific form of the model; and, second, to generalize it so as to take account of important features of the real world that it leaves out. I have already mentioned the question of adaptation and adjustment rates. But we should not abandon the idea of a coherent and manageable approximation to reality or we shall fail to obtain even an approximate picture of the structure of demand.

4. SOME RESULTS OF THE MODEL

I wish that I could now give a complete set of results obtained from the model I have described. Unfortunately I cannot because the series of analyses I have outlined are not yet completed. So far we have only analysed eight main groups; we have made no subgroup analyses and we have made no systematic comparison with budgets.

The data we have used relate to expenditure per head on eight commodity groups and the corresponding price index-numbers measured over the years 1900 to 1960. In estimating b and c we left out the years 1914 through 1919 and 1940 through 1947 because of the abnormal conditions of war periods. This means that we have 376 observations from which to determine 30 independent parameters in the case of the model with linear trends and 45 in the case of the model with quadratic trends. The method of fitting is the one described in the preceding section with θI in place of $i' e_\theta^* I$ and $\hat{e}_\theta^*$ in the case of the linear version and with additional terms of the form $\theta^2 b^{***}$ and $\theta^2 c^{***}$ in the case of the quadratic version.

The estimates of the parameters are shown in the following two tables and the goodness of fit of the two models is indicated in the double page diagram on pp. 200–201.

The first table compares the components of b, the second those of c. The letters L and Q denote respectively the linear and the quadratic model.

In these tables $\theta = 0$ in 1960 and so b^* and c^* show the estimates of b and c in the year. All these estimates showed considerable trends. For example, in the linear model the proportion of uncommitted expenditure spent on food fell by 0·39 percentage points in each year until in 1960 it was as low as 8·05 per cent. At the same time committed purchases per head of food measured in 1938 prices rose each year by £0·332 until in 1960 they reached

TABLE 1

	b^*		b^{**}		b^{***}	
	L	Q	L	Q	L	Q
Food	0·0805	0·0948	−0·00390	0·00039	—	0·000086
Clothing	0·1569	0·1813	0·00210	−0·00008	—	−0·000035
Household	0·2263	0·1724	0·00210	−0·00961	—	−0·000137
Communications	0·0023	−0·0018	−0·00040	0·00010	—	0·000002
Transport	0·2342	0·2588	0·00160	0·01080	—	0·000117
Drink and tobacco	0·0956	0·0618	−0·00190	−0·00161	—	−0·000038
Entertainment	−0·0143	−0·0018	−0·00080	0·00236	—	0·000035
Other	0·2186	0·2346	0·00120	−0·00236	—	−0·000029
Total	1·0001	1·0001	0·00000	−0·00001	—	0·000001

TABLE 2

	c^*		c^{**}		c^{***}	
	L	Q	L	Q	L	Q
Food	33·73	35·06	0·3320	0·2575	—	−0·000518
Clothing	9·26	12·25	0·0430	0·0557	—	−0·000544
Household	24·59	28·63	0·0540	−0·0383	—	−0·004702
Communications	1·13	1·15	0·0260	0·0358	—	0·000366
Transport	8·97	14·62	0·1650	0·3979	—	0·003135
Drink and tobacco	10·20	11·56	0·0580	0·1572	—	0·003829
Entertainment	5·05	4·86	0·0750	0·1035	—	0·000778
Other	9·99	14·15	0·0640	−0·0229	—	−0·002401
Total	102·92	122·28	0·9170	0·9464	—	−0·000057

£33·73. In 1960 uncommitted expenditure according to the linear model was £38·32 per head and committed expenditure on food at 1960 prices was £93·1 per head. Thus, according to the linear model, food expenditure per head in 1960 should have been £[93·1+(38·32 × 0·0805)] = 96·2. The corresponding figure from the quadratic model is £95·3. The observed value is £95·8. In each of these cases the error is about one half of one per cent.

The diagram overleaf shows that both models perform well. In many cases, e.g. throughout the period for food and in the post-war years for all the series, they show very similar results. In other cases, where they differ, the quadratic model is usually the better. For example, it shows to advantage in reproducing the inter-war series for clothing and household almost exactly and gives a much better reproduction of the earlier part of the series for transport and communications. The rather flat series for transport obtained for the inter-war period from the quadratic model is probably due to the substantial element of net investment in the purchases of cars in that period. Since neither version of the general model given here takes any account of consumers' problems of adaptation, each is likely to reproduce the underlying movement of consumption, that is purchases minus net investment. The other case in which this kind of effect might be important is the durable component of the category household. But here the net investment component of purchases is smaller than in the case of cars and the category itself, which includes expenditure on rent and fuel, is very large.

Although the two war periods were left out in calculating b and c, the actual and estimated series for these periods are shown in the diagram. On the whole they show what one might expect. For example, during the second world war the average consumer systematically devoted less of his expenditure to food and clothing and more of it to drink and tobacco than he would have done in normal circumstances.

It will be noticed that in the post-war years the fit of both models is uniformly good. This is not surprising because it is the sum of squares of all the absolute discrepancies that is being minimized by the statistical procedure, and in the post-war period all series were relatively high because prices were high compared with most earlier periods. What is perhaps more interesting is that the models also fit reasonably well in the early part of the century when prices were very much lower and also in the case of the very small groups, communications and entertainment.

As I have said, we have not yet made a systematic comparison between the total expenditure elasticities derived from the model and those derived from budgets. So far we have only looked into two cases, food and clothing, in both of which the alternative estimates are in fairly close agreement. For food, the linear model gives estimates of this elasticity of 1·0 for 1900, 0·6 for 1938 and 0·3 for 1960; the quadratic model gives corresponding estimates of 1·1, 0·4 and 0·3; budgets are not available in sufficient detail for the earlier part of the period but give estimates of 0·6 for 1938 [1, 168] and 0·3 for 1960 [180], in complete agreement with the linear model. For clothing, the linear model gives estimates of 0·3 for 1900, 1·1 for 1938 and 1·4 for 1953; the quadratic model gives corresponding estimates of 0·7, 1·6

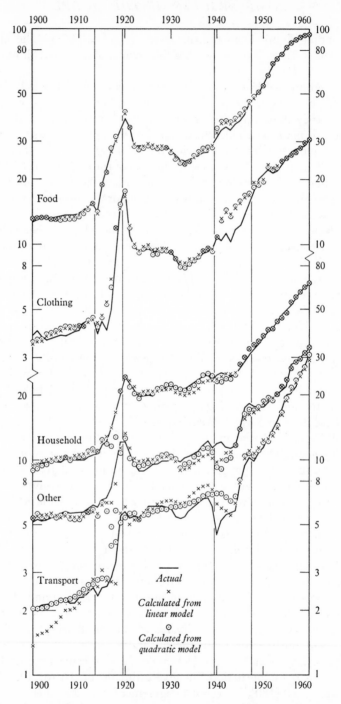

Consumers' Expenditure in Britain, 1900–1960

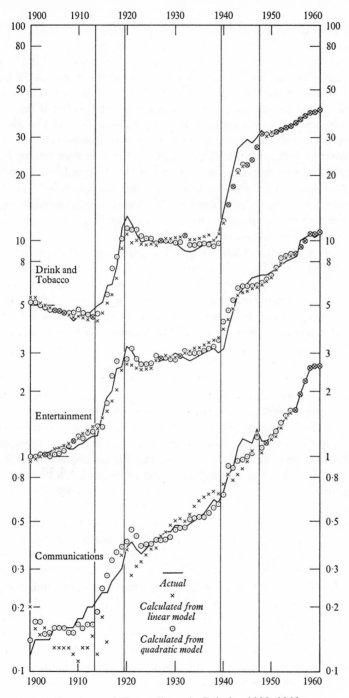

Consumers' Expenditure in Britain, 1900–1960

and 1·7; budgets give estimates of 1·1 for 1938 [127] and 1·4 for 1953 [57], again in complete agreement with the linear model.

It is hard to say why the linear model should have appeared the better one in the comparisons. It may be that the budget estimates are not the best that could be made; it may be that the curvilinearity introduced in the quadratic model is not of the appropriate kind. Reasons for this belief have already been given though no calculations have yet been made with the third time-series model or with the model in which budgets and time series are combined for estimation purposes. In the present case it is the parameter b that is important and it is, perhaps, significant that, in the quadratic model, $\min b$ for food occurs in 1958 and that $\max b$ for clothing occurs in 1959. Thus in both cases, over the observation period, there is a considerable degree of curvature which has already changed direction. Certainly we might be doubtful of such results in making projections.

Another curious feature of the quadratic model is that in almost all years uncommitted expenditure is negative. This is in contrast with the result obtained from the linear model in which uncommitted expenditure is positive except during wars and the years immediately following them. However, even with this model uncommitted expenditure is not very big; in the inter-war and post-war periods it fluctuated between 10 and 15 per cent of total expenditure.

This result of the linear model seems to me to be much what one might expect. The average consumer has a great many commitments, whether they are legal commitments or not, and can only make marginal adjustments to his expenditure pattern. But over time he tends to get richer and as he does his conception of his commitments rises. As a consequence he can still only make marginal adjustments. In war and immediate post-war years the average consumer cannot maintain his conception of his standard of living and, in one way or another, is forced to cut back his expenditure. As conditions improve a margin of uncommitted expenditure reasserts itself.

The quadratic model goes further than this and suggests that the average consumers' conception of his standard of living always outruns the money he has to spend and so he is always preoccupied not with what to buy but with what not to buy. In a society in which emulation plays an important part and which is not divided by sumptuary laws into non-competing groups, even this picture may not seem unduly far-fetched. But if we accept it, the whole theoretical interpretation of the model needs to be reconsidered since, as we can see from (8), a negative value of uncommitted expenditure would cause the own-elasticities of substitution, $s_{\beta\beta}$, to be positive. The position would be still worse if the commitments were so high that the term in round brackets on the right-hand side of (4) became negative, since in this case the Marshallian demand curves would get into the wrong quadrant.

The Slutsky condition (8) is an equilibrium condition, and we might argue that the average consumer cannot be in equilibrium if his uncommitted expenditure is negative. With the quadratic model this makes the theory somewhat remote from reality, unless, as is possible, we have so far been able to make only a poor approximation to the true values of the parameters in this model.

5. CONCLUSIONS

My conclusions from the work discussed in this paper can be summarized as follows.

(1) The kind of demand model I have described is capable of a sensible interpretation and gives a reasonably good representation of the past.

(2) This model is capable of considerable generalization and of making very full use of past experience.

(3) In it, the influences of income and prices are introduced in a simple and approximate way. The important innovation, on which I have concentrated in this paper, is the allowance for changing responses. This is done by allowing the parameters, which have a clear economic interpretation, to change over time.

(4) An outline of our practical experience with the model shows that it is promising but that there is still room for experiment and improvement. Two methods of improvement were suggested: the introduction of more sophisticated trends in the parameters and the combination of time series and budget studies.

(5) The model is hierarchical: it deals first with main groups, then with subgroups of main groups and then with subgroups of subgroups. The thought here is that main groups are relatively little affected by changes either from the side of supply or of demand which cannot be represented by simple systematic changes in the parameters. As the subgroups become smaller the effect of these changes become more important until at some point in the subdivision of commodities the usefulness of the model is likely to be exhausted. This is as it should be; it is easier to formulate relationships to describe broad categories than to describe narrow ones. It is helpful to be able to see how far we can go with any proposed formulation.

(6) In all our work on economic growth we are concerned to produce the best picture we can of the future based on the changing relationships of the past. By this means we expect to be able to provide a worthwhile basis of discussion with those engaged in the different branches of economic activity. From such discussions we hope that a plan will finally emerge which is realistic as well as consistent, and which reaches a reasonable compromise between individual and social aims.

XV

PRIVATE SAVING IN BRITAIN, PAST, PRESENT AND FUTURE

1. INTRODUCTION

Nearly two years ago, Mr D. A. Rowe and I published a model designed to explain the course of personal saving and showed that it gave a good account of the variations that took place over the 1950's [167]. In that paper we predicted a considerable fall in personal saving between 1961 and 1962 and, in the event, we were fairly near the mark. In this paper I shall go through the calculations again with the revised statistics now available, extend the original model to allow for differential behaviour on the part of wage and profit earners and make a forecast for 1964. I shall also apply the model to companies, so as to give a picture of private saving as a whole, and examine the saving ratios that might be expected if the income of persons and companies grew steadily at, for example, the N.E.D.C. rate of 4 per cent.

This paper is the first fruit of the extension into the financial sphere of the work on British economic growth undertaken by a group of us at the Department of Applied Economics in Cambridge. In a recent paper [159] we made a preliminary estimate of the saving ratio that would be needed if conditions similar to those envisaged by the N.E.D.C. were realized by 1970. In the final part of this paper, I shall try to discover whether, under the conditions assumed, the private sector of the economy would save the required amount if it continued to behave in respect to saving as it appears to have behaved over the years since 1949.

2. A MODEL OF SAVING

In this section I shall describe the model of saving behaviour as best I can in words; the reader who feels more at home with algebra will find a mathematical statement in appendix 1.

The main idea is that saving is responsive to wealth as well as to income. Other things being equal, an increase in income will lead to an increase in saving but an increase in wealth will lead to a decrease in saving. Put another way, spending, which is equal to income minus saving, responds positively to both income and wealth. Other things being equal, an increase in income will be divided in a certain way between spending and saving; and again, other things being equal, an increase in wealth will lead to more spending and so to less saving.

The second idea is that an increase in income may be regarded on the one hand as permanent, or likely to continue, or, on the other hand, as transient, or unlikely to continue. It seems probable that an increase in income will be differently divided between additional spending and saving according as it is believed to be permanent or transient. People are unlikely, for example, to use a windfall gain of income to increase their normal expenditure since, if they do, they will get into spending habits which they cannot expect to maintain. On the other hand they may well use a windfall gain of income for a spending spree and, in particular, for the purchase of durable goods which could be maintained but not acquired out of permanent income.

In trying to give effect to these two main ideas, we must realize that although we have regular statistics of disposable income, spending and saving we do not have regular statistics of wealth; also, we have no direct means of dividing any of these variables into a permanent and transient component. We can, however, meet these difficulties as follows.

We begin by dividing income, wealth and expenditure notionally into two, a permanent part and a transient part. We assume in the case of income and wealth that this year's permanent part is a weighted average of this year's total and last year's permanent part. This is the same as saying that the increase in the permanent components from last year to this year are a constant proportion of the excess of this year's total over last year's permanent component. We now say that the permanent part of expenditure is a homogeneous linear function of the permanent components of income and of wealth and that the transient part of expenditure is proportional to the transient component of income. These assumptions are formulated in equations (8) through (12) of appendix 1, where it is shown that, as a consequence, expenditure and therefore saving in any year can be expressed as a homogeneous linear function of the income of that year, the wealth at the beginning of that year and the income and the expenditure of the preceding year. These results are set out in equations (15) and (16) from which an expression for the saving ratio is immediately derived in (17). In this way we clear the unobservable permanent and transient components out of the final equations. In applying the model we express all the variables at constant consumer prices and define wealth as accumulated saving at constant prices. Thus if we have an estimate of wealth at the beginning of one year we can obtain estimates for all other years from our estimates of saving.

I show in appendix 1 how the simple model can be extended in various ways. The most interesting extension relates to the subdivision of income. Economists frequently assume that the greater part of saving is made out of property income, little or none being made out of wages or small transfers. We shall see in the following section that this is an over-simplification although personal saving responds more to a change in property income than to a change in income of other forms.

Another point which must be recognized in applying the model is that it depends on the time units used; a quarterly version cannot, in general, be aggregated to give an annual version of the same form. In the following two sections I shall show that both versions work out well and that they are consistent in the one point they have in common.

Finally, since saving is the rate of change of wealth, the model leads to a differential equation from which we can work out the equilibrium values of the wealth–income and saving–income ratios which would be realized under conditions of a constant rate of growth of income.

Thus the essence of the theory of saving presented in this paper is that economic units have the notion of a desirable relationship between their income and their wealth which they are constantly trying to realize. This relationship is such that the faster they grow in terms of income the higher is their saving–income ratio and the lower is their wealth–income ratio. If individuals have what they consider an excess of wealth in relation to income they will tend to reduce their saving and so allow their income to rise relative to their wealth. In the opposite case they will tend to increase their saving and so enable their wealth to rise relative to their income.

These ideas can be applied to companies as well as to persons, but in this case it is perhaps more natural to speak not of an intended wealth–income ratio but of its reciprocal, an intended rate of return. On this basis company saving and spending behaviour is seen as determined by the effort to realize a given rate of return. Spending, here, means of course spending on dividends.

3. ANNUAL MODELS OF PERSONAL SAVING

The variables used in these models are taken mainly from the *Blue Books* on national income and expenditure [176] and are set out in table 1. They are expressed at 1958 prices and are defined as follows.

Income, denoted by μ, is disposable personal income, that is income as defined in the *Blue Book* less: (i) personal provisions for depreciation and stock appreciation; (ii) payments of taxes on income and of national insurance and health contributions; and (iii) net remittances abroad. It is equal to personal spending plus personal saving as defined below. The two sub-divisions of income, denoted by μ_a and μ_b relate respectively to income from employment and transfers and to all other forms of personal income.

Spending, denoted by ϵ, is consumers' expenditure including expenditure on all forms of consumers' durables except dwellings.

Saving, denoted by σ, is the balance of personal saving, as defined in the *Blue Book* plus additions to tax reserves and less personal provisions for depreciation and stock appreciation.

Wealth, denoted by ω, is personal net worth. For the purpose of this paper a provisional figure of £55 500 million at 1958 prices was given to me by my colleague Mr J. R. S. Revell for the beginning of 1961. The remaining figures are obtained by additions and subtractions of annual saving.

Government measures to discourage spending, denoted by ξ, are represented by the percentage downpayment on the hire-purchase of radio and electrical goods. This indicator was on the whole closely related to other measures designed to affect spending, such as ease of credit.

A number of calculations were made using the data in table 1. The first was based on equation (23) of appendix 1 and gave the following results.

$$\frac{\sigma}{\mu} = \frac{0\cdot413}{(0\cdot084)} - \frac{0\cdot0464}{(0\cdot0168)}\frac{\omega}{\mu} + \frac{0\cdot130}{(0\cdot157)}\frac{E^{-1}\mu}{\mu} - \frac{0\cdot366}{(0\cdot210)}\frac{E^{-1}\epsilon}{\mu} + \frac{7\cdot24}{(2\cdot06)}\frac{\Delta^*\xi}{\mu} \quad (1)$$

In this case the model accounts for just over 97 per cent of the variance of the saving ratio. The numbers in brackets indicate the standard errors of the regression coefficients; thus, we might write the constant term in (1) as $0\cdot413 \pm 0\cdot084$. There is no evidence of serial correlation in the residuals of this equation; using Durbin and Watson's measure, $d = 1\cdot96$.

If we multiply all the terms in (1) by μ, we obtain an equation for σ; in this form the coefficients are perhaps most easily interpreted. Thus a rise of £1 in income is accompanied other things being equal by a rise of £0·413 in saving. A rise of £1 in initial wealth is accompanied other things being equal by a fall of £0·0464 in saving. A rise of £1 in last year's income is accompanied other things being equal by a rise of £0·130 in this year's saving whereas a rise of £1 in last year's expenditure is accompanied by a fall of £0·366 in this year's saving. Finally, a change from last year in government measures to discourage spending by one unit is accompanied by a rise of £7·24 million in personal saving.

In (1), each coefficient is larger than its standard error with one exception: the coefficient of the term in $E^{-1}\mu/\mu$. We might therefore put this coefficient equal to zero, and this is equivalent to saying that people do not spend any part of transient income. With this omission the equation becomes.

$$\frac{\sigma}{\mu} = \frac{0\cdot439}{(0\cdot077)} - \frac{0\cdot0570}{(0\cdot0108)}\frac{\omega}{\mu} - \frac{0\cdot222}{(0\cdot116)}\frac{E^{-1}\epsilon}{\mu} + \frac{6\cdot93}{(1\cdot99)}\frac{\Delta^*\xi}{\mu} \quad (2)$$

The goodness of fit of this model is hardly inferior to that of (1). The only striking change is that the coefficient of $E^{-1}\epsilon/\mu$ is reduced. In fact last year's income and last year's expenditure are closely related and so the omission of the term in last year's income results in a coefficient for the term in last year's expenditure which is approximately equal to the sum of the co-efficients of these terms, $0\cdot130 - 0\cdot366 = -0\cdot236$, in (1). In each case the standard errors in (2) are less than those in (1) and, in my opinion, (2) is to be preferred to (1) as an expression for the saving ratio although it must be admitted that there is very little to choose between the two equations.

A third equation was fitted to the data making use of the two income variables, μ_a and μ_b, and of equation (24) in appendix 1. Here, again, it was found that the coefficients of the terms involving last year's income were less than their standard errors. The version corresponding to (2) above is as follows:

$$\frac{\sigma}{\mu} = \frac{0\cdot395}{(0\cdot095)}\frac{\mu_a}{\mu} + \frac{0\cdot573}{(0\cdot181)}\frac{\mu_b}{\mu} - \frac{0\cdot0631}{(0\cdot0133)}\frac{\omega}{\mu} - \frac{0\cdot183}{(0\cdot127)}\frac{E^{-1}\epsilon}{\mu} + \frac{7\cdot22}{(2\cdot06)}\frac{\Delta^*\xi}{\mu} \quad (3)$$

This equation is not very different from the preceding one. We find, as we expected, that the response of saving to a change in income from employment and small transfers is less than its response to a change in other forms of income. The difference is, however, less than might have been supposed.

TABLE A   STRUCTURAL COEFFICIENTS IN ANNUAL MODELS OF PERSONAL SAVING

	Equation number		
	(1)	(2)	(3)
α_1	0·0732	0·0732	0·0773
β_1	0·721	0·721	—
β_{1a}	—	—	0·741
β_{1b}	—	—	0·523
β_2	0·355	—	—
λ	0·634	0·778	0·817
δ	−19·8	−31·3	−39·3
R^2	0·971	0·969	0·971
d	1·96	1·76	2·03

As explained in (28) through (31) of appendix 1, we can derive the structural coefficients of the model from the estimated regression coefficients .This is done, for the three equations just given, in table A above. To these values I have added two technical measures: R^2, the square of the coefficient of multiple correlation; and d, Durbin and Watson's test statistic for serial correlation.

These estimates can be interpreted as follows. The coefficient α_1 represents the response of spending, other things being equal, to a change in the permanent component of wealth: an increase of £1 in this component of wealth leads to an increase of about 1s. 6d. in spending and therefore to a fall of the same amount in saving. The coefficient β_1 represents the response of spending to a change in the permanent component of income: an increase of £1 in this component of income leads to an increase of about 14s. 6d. in spending and therefore to an increase of about 5s. 6d. in saving. In the column for equation (3) we see β_1 divided into two parts: β_{1a}, associated with income from employment and transfers, and β_{2a}, associated with other forms of income. Whereas an increase of £1 in the permanent component of the first kind of income leads to an increase of about 15s. in spending and 5s. in saving, a similar increase in the same component of the second kind of income leads to an increase of about 10s. 6d. in spending and 9s. 6d. in saving. Since the first kind of income is numerically the more important, we find the average response in columns (1) and (2) heavily weighted in its favour.

The coefficient β_2 represents the response of spending, other things being equal, to a change in the transient component of income. We have already seen that the estimate, which comes from the term in $E^{-1}\mu/\mu$, is highly uncertain. It is nevertheless quite large; according to the figure in

DIAGRAM 1   PERSONAL SAVING RATIO: BRITAIN, 1949–62

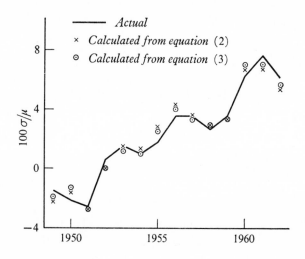

the first column an increase of £1 in this component of income leads to an increase of about 7s. in spending and so of 13s. in saving. But with the information we possess we cannot be sure about these numbers. By adjusting our ideas about the proportion of the difference between this year's income (and wealth) and last year's permanent component which is regarded as permanent, that is by adjusting λ, we can always put $\beta_2 = 0$ if we wish. A change in λ is also accompanied by a change in δ, which measures the effect on spending of a change in government measures from the level previously regarded as normal.

The extent to which the three equations account for variations in the personal saving ratio is shown in table 1 of appendix 2. A graphical comparison of the actual course of events and the estimates from equations (2) and (3) is shown in diagram 1.

4. A QUARTERLY MODEL OF PERSONAL GROSS SAVING

A consistent set of quarterly data is available in *Economic Trends* [175]. In this source, income and saving are defined before deducting provisions for depreciation and stock appreciation. In [167], Mr Rowe and I found that, in a quarterly model, the income of the previous quarter was a significant variable whereas the change in government restrictive measures was not. This experience is reflected in the results set out in equation (4) below. In

DIAGRAM 2   PERSONAL GROSS SAVING RATIO: BRITAIN, 1955–64

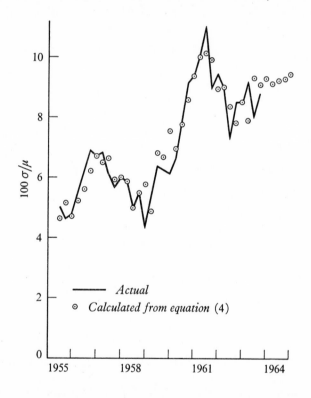

spite of the fact that income and expenditure are now gross, I shall use the same symbols as before.

For the thirty-two quarters, from the second quarter of 1955 through the first quarter of 1963, the following equation was obtained:

$$\frac{\sigma}{\mu} = \frac{0{\cdot}472}{(0{\cdot}121)} - \frac{0{\cdot}0130}{(0{\cdot}0060)}\frac{\omega}{\mu} + \frac{0{\cdot}394}{(0{\cdot}111)}\frac{E^{-1}\mu}{\mu} - \frac{0{\cdot}684}{(0{\cdot}145)}\frac{E^{-1}\epsilon}{\mu} \tag{4}$$

The constants in this equation can be interpreted in precisely the same way as those of the first three equations which were discussed in detail in the preceding section. As already mentioned and as demonstrated in equation (26) of appendix 1, a quarterly model cannot be aggregated into an annual model of the same form. The point of contact of the two models lies in the absolute value of the coefficient of the term in $E^{-1}\epsilon/\mu$; the fourth power of the value in the quarterly model should be equal to the value in the annual model. It is interesting to note that $(0\cdot684)^4 = 0\cdot219$ which can be compared with the value $0\cdot222$ in (2) above.

The structural coefficients and other measures for (4), corresponding to the values in table A, are as follows: $\alpha_1 = 0\cdot0411$; $\beta_1 = 0\cdot411$; $\beta_2 = 0\cdot583$; $\lambda = 0\cdot316$; $R^2 = 0\cdot893$; and $d = 2\cdot17$.

The extent to which this equation accounts for variations in the gross saving ratio is shown in table 2 of appendix 2. A graphical comparison with the actual course of events is shown in diagram 2.

5. SHORT-TERM FORECASTS OF PERSONAL GROSS SAVING

In this section I shall examine the outcome of the prediction made in [167] for 1962 and make a new prediction for 1964. All these calculations are made with the quarterly model and so relate to gross saving.

(a) *The 1962 prediction.* This calculation was made early in 1962 on the basis of the quarterly data then available which extended from the first quarter of 1955 through the last quarter of 1961. The constants used were estimated over the period of twenty-four quarters ending with the first quarter of 1961.

We have seen in (4) that gross saving in any quarter depends on four variables: the gross disposable income of the quarter, the wealth at the beginning of the quarter and the gross disposable income and the expenditure of the preceding quarter. Thus if we guess the gross disposable income in the first quarter of 1962 we can estimate gross saving in that quarter and therefore wealth at the beginning of the following quarter. By guessing gross disposable income in successive quarters we can therefore build up a succession of quarterly forecasts, and we can add these together to give a forecast for the year.

The forecast made in 1962 was in fact quite accurate, within 3 per cent of the latest figure for personal gross saving in 1962 given in [167] when adjusted to 1958 prices. The original forecast was made at 1954 prices, but for comparative purposes it is revalued at 1958 prices in table B below.

The revised estimates relate to the latest figures now available in [175; no. 120] and to calculated values based on a set of constants similar to those in (4) above but estimated from data for the period of twenty-four quarters from the second quarter of 1955 through the first quarter of 1961. The figure of £1595 million was reached by the same method as the original figure of £1601 million except that the actual series for gross disposable income was used in place of the original guesses. A comparison of the last two columns of table B provides, therefore, a test of the model over this period. The new

15

calculated value for 1962 is very like the original one: rather less than 3 per cent above the actual figure.

In the original paper [167], this forecast was presented in the worst possible way, namely as a change on the actual level of 1961. As we can see from table B, the peak value of 1961 has been substantially reduced in the latest official revisions.

TABLE B   PERSONAL GROSS SAVING IN 1961 AND 1962

(1958 £ million)

	Original estimates		Revised estimates	
	Actual	Calculated	Actual	Calculated
1961	2032	1962	1826	1799
1962	—	1601	1555	1595

(*b*) *A prediction for* 1964. A similar calculation for 1964 is set out in table C below. In making this forecast I have used the data given in [175; no. 120] supplemented by figures for the third quarter of 1963 given in [133; no. 48]. I have assumed that gross disposable income at 1958 prices will increase at approximately 4 per cent a year from 1962 to 1963 and again from 1963 to

TABLE C   PERSONAL GROSS SAVING IN 1963 AND 1964

(1958 £ million)

	Actual	Calculated
1963	1701*	1700
1964	—	1847

*Last quarter calculated

1964. I have made the first of these increases rather smaller than the second on account of the very cold winter at the beginning of 1963. I have made use of the coefficients shown in (4) above.

Thus, on the assumptions made about gross disposable income, the model suggests a fairly steady recovery of personal gross saving at 1958 prices from the low level of 1962 to a figure somewhat above the peak of 1961. Comparing tables B and C we see a fall in 1961–62 of £271 million followed by rises in 1962–63 and 1963–64 of £146 million in each case.

6. ANNUAL MODELS OF COMPANY SAVING

The original paper [167] was restricted to the analysis of personal saving. In this section I shall analyse company saving using the model set out in equation (20) of appendix 1. I shall analyse first the data available in [176] relating to all companies and second the data available in [175; no. 102] relating to quoted companies in manufacturing and distribution.

(a) *All companies.* The information given in the *Blue Book* [176] makes possible a number of alternative definitions of income and saving. The ones I shall use here can be described as follows.

Income is defined after payment of all interest, profits due abroad, provisions for depreciation, taxes paid abroad and British taxes on retained income. It is therefore equal to dividend payments before deduction of tax plus saving as defined below.

Spending is defined as dividend payments before British taxes on income in respect of preference and ordinary shares.

Saving is defined as undistributed income, including stock appreciation but after provisions for depreciation, plus all additions to dividend and tax reserves.

Wealth is defined as company net worth. Pending the completion of Mr Revell's researches on this subject, a figure of £30000 million at 1958 prices was taken for the beginning of 1960 and the remaining figures were obtained by additions and subtractions of annual saving.

The result of fitting equation (20) to the data for 1949 through 1962 given in table 3 of appendix 2 was as follows:

$$\frac{\sigma}{\mu} = \frac{0{\cdot}901}{(0{\cdot}041)} - \frac{0{\cdot}00752}{(0{\cdot}00339)}\frac{\omega}{\mu} + \frac{0{\cdot}0584}{(0{\cdot}0418)}\frac{E^{-1}\mu}{\mu} - \frac{0{\cdot}844}{(0{\cdot}126)}\frac{E^{-1}\epsilon}{\mu} + \frac{116}{(88)}\frac{1}{\mu} \quad (5)$$

The structural coefficients and other measures for (5) are: $\alpha_1 = 0{\cdot}0484$; $\beta_1 = 0{\cdot}261$; $\beta_2 = 0{\cdot}0691$; $\lambda = 0{\cdot}156$; $\gamma = -138$; $R^2 = 0{\cdot}932$; and $d = 1{\cdot}34$.

(b) *Quoted companies in manufacturing and distribution.* The information given in *Economic Trends* [175; no. 102] could again give rise to alternative definitions.

Income is defined as disposable income less share due to minority interest.

Spending is defined as ordinary and preference dividends (net).

Saving is defined as balance retained in reserves.

Wealth is defined as shareholders' interest in total net assets. These figures are estimated directly for each year.

The result of fitting equation (20) to the data for 1950 through 1960 given in table 4 of appendix 2 was as follows:

$$\frac{\sigma}{\mu} = \frac{0{\cdot}744}{(0{\cdot}079)} - \frac{0{\cdot}0461}{(0{\cdot}0134)}\frac{\omega}{\mu} - \frac{0{\cdot}0139}{(0{\cdot}1084)}\frac{E^{-1}\mu}{\mu} + \frac{0{\cdot}297}{(0{\cdot}486)}\frac{E^{-1}\epsilon}{\mu} + \frac{186}{(63)}\frac{1}{\mu} \quad (6)$$

Here we see that the coefficient of the term in $E^{-1}\mu/\mu$ is much smaller than its standard error. If we leave out this term, we obtain

$$\frac{\sigma}{\mu} = \frac{0 \cdot 749}{(0 \cdot 063)} - \frac{0 \cdot 0453}{(0 \cdot 0109)}\frac{\omega}{\mu} + \frac{0 \cdot 246}{(0 \cdot 255)}\frac{E^{-1}\epsilon}{\mu} + \frac{180}{(35)}\frac{1}{\mu} \tag{7}$$

The curious feature of (6) and (7) is that in each case the coefficient of $E^{-1}\epsilon/\mu$ is positive, implying that more than the whole of an increase in

DIAGRAM 3   COMPANY SAVING RATIOS: BRITAIN, 1949–62

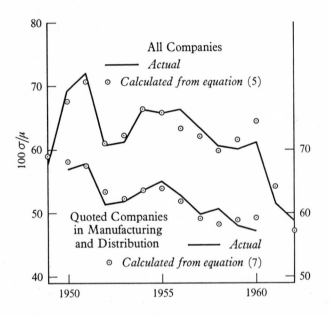

income above the level regarded as permanent in the previous year is considered permanent. I have no satisfactory explanation to offer for this apparently extreme form of expectation, but we must note that the estimates are very unreliable.

The structural coefficients and other measures for (7) are as follows: $\alpha_1 = 0 \cdot 0364$; $\beta_1 = 0 \cdot 201$; $\beta_2 = 0$; $\lambda = 1 \cdot 246$; $\gamma = -144$; $R^2 = 0 \cdot 856$; and $d = 1 \cdot 89$. The values for (6) are very similar.

Thus we see that the structural coefficients of (5) and (7) are not wildly different except in the case of λ.

The extent to which the equations of this section account for variations in company saving ratios is shown in tables 3 and 4 and illustrated graphically in diagram 3.

7. RATIOS AND STEADY GROWTH

The model of saving behaviour which I have been describing and illustrating implies that if income grows at a constant rate the wealth–income and saving–income ratios will tend to equilibrium values. This result is derived in equations (32) through (40) of appendix 1. Thus, from (37) and (40) we see that each ratio contains a transient term which would be equal to zero if the initial wealth–income ratio happened to take the equilibrium

DIAGRAM 4   PERSONAL SAVING RATIO

Steady-state values for different growth rates of disposable personal income

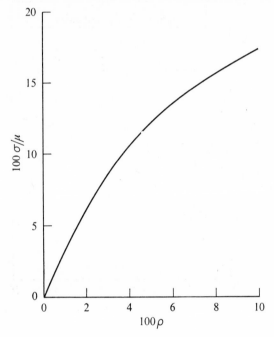

value for the growth rate considered but which will tend to zero in any case; and a steady-state term which is a constant for the saving ratio and also for the wealth ratio in the homogeneous case ($\gamma = 0$), and tends to a constant for the wealth ratio in the general case ($\gamma \neq 0$).

The steady-state component of the equilibrium saving ratio is shown as a function of the growth rate of income in table 5 and illustrated for persons in diagram 4 and for companies in diagram 5. For persons I have based the calculations on equation (2) and for companies on equations (5) and (7). We can see that the steady-state ratios are much lower for persons than for companies and that the two superficially different equations for companies give fairly similar values for the equilibrium ratios.

Let us now make use of these calculations to compare the steady-state values of the saving ratios with the actual values observed in recent years. With a steady growth rate of 2 per cent we should expect persons to save on the average about 6 per cent of their income, and this in fact was the figure reached in the early years of the present decade for the first time since the war. If the N.E.D.C. growth rate of 4 per cent were brought about we should expect persons eventually to save about 10 per cent of their income.

DIAGRAM 5   COMPANY SAVING RATIO
Steady-state values for different growth rates of company disposable income

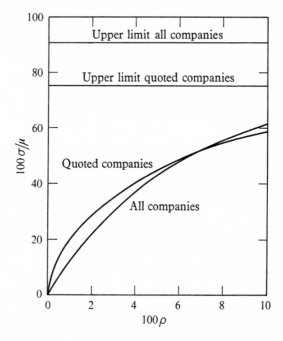

So, if personal income were to grow steadily at the N.E.D.C. rate we should expect more of it to be saved than is being saved at present.

When we turn to companies we see exactly the opposite picture. Compared with the steady-state values in the growth range 2 to 4 per cent, company saving ratios in the 1950's were extremely high. True, these ratios have shown a downward tendency for nearly a decade but by 1962 the figure for all companies was about 49 per cent compared with a steady-state value of under 40 per cent for steady growth at the rate of 4 per cent. So, if company income were to grow steadily at the N.E.D.C. rate we should expect less of it to be saved than is being saved at present.

The final question I shall consider is: what happens if we put these divergent tendencies together and compare the result with what the

N.E.D.C. growth rate would seem to demand? In a recent preliminary study of the British economy in 1970 [159], a study which implied a growth rate of British production of exactly 4 per cent, we suggested that British saving would have to rise from the 10·1 per cent of income it was in 1960 to about 11·5 per cent of income in 1970. This increase seemed to us at the time by no means impossible; let us now compare it with the results of this paper.

In order to make the comparison I propose to set out in a simplified form the capital transactions account for the British economy in each of the three

TABLE D   SAVING, INVESTMENT AND INCOME,
ACTUAL AND HYPOTHETICAL

(£ *million and percentages*)

	1960		1961		1962	
	Actual	Hypo-thetical	Actual	Hypo-thetical	Actual	Hypo-thetical
Saving						
Persons	1 062	1 816	1 415	1 938	1 166	2 007
Companies	1 709	1 044	1 233	890	1 106	861
Government	−42	−42	−2	−2	358	358
Residual error	−303	−303	55	55	−136	−136
Investment	2 426	2 515	2 701	2 881	2 494	3 090
Income	23 377	23 377	24 911	24 911	25 910	25 910
Ratios						
Total saving to income	10·4	10·8	10·8	11·6	9·6	11·9
Private saving to income	11·9	12·2	10·6	11·4	8·8	11·1

years, 1960, 1961 and 1962. I shall show this account first as it actually was and second as it would have been if persons and companies had saved according to the steady-state ratios appropriate to a growth rate of income of 4 per cent. The results are set out in table D.

The purpose of setting out these figures for three years is to enable some allowance to be made for varying conditions and for the effect of fluctuations in government saving and the residual error. Let us, however, first consider the conclusions and then come to the peculiarities of the statistics. If we look at the penultimate entry in the first column, we see that the actual ratio of total saving to income has been revised upwards from the 10·1 per cent

I mentioned earlier to 10·4 per cent. If we look at the first ratio in each of the hypothetical columns, we see that they form a rising series, 10·8, 11·6 and 11·9 with an average of 11·4. These ratios show the percentage of income that would have been saved in each year if government saving and the residual error had been what they were but personal and company saving had taken on the steady-state proportions appropriate to a 4 per cent growth of income. The second set of ratios in the hypothetical columns show the relationship of private saving to total income under the hypothetical conditions assumed. These ratios form a falling series, 12·2, 11·4 and 11·1 with an average of 11·6.

Thus the conclusion we come to is that the supply of saving that might be expected to accompany a 4 per cent growth rate would be equal to the demand although the relative importance of the principal sources of supply, persons and companies, would be completely changed. I need hardly emphasize the tentative nature of this conclusion resting, as it does, on analyses of the supply and demand of saving which need to be checked and tested further.

I cannot end this section without a word about the figures in table D. The actual figures are taken from [176]. The figures for saving should add up to investment but with the sources of information at present available there is a residual error which, following the usual practice, I have treated as part of saving. Personal saving in table D is defined as in section 3; but the definition of company saving in the table differs from the definition in section 5 by the exclusion of stock appreciation at its actual value in the different years. Government saving includes the saving of public corporations as well as that of the central government and local authorities. Income is the national income plus net indirect taxes, the concept used in calculating the required saving ratio in [159].

In table 5, values are given of the equilibrium wealth–income ratios calculated from the middle term on the right-hand side of equation (37). These ratios invite comparison with similar ratios derived from the requirements of production. But this is another subject, and I shall not try to develop it in this paper.

8. CONCLUSIONS

I hope I have made clear the conclusions of the different sections; but, as this is a long paper, it may be useful if I try to summarize them.

(1) Following [167], I have described a model of saving behaviour which makes use of the idea of an equilibrium ratio of wealth to income and of a distinction between permanent and transient components of income and wealth.

(2) An application of the model to annual data on personal saving shows that it works well and gives an intelligible first approximation to personal

saving behaviour. We have seen that the marginal propensity to save out of wages and small transfers is less than the marginal propensity to save out of other forms of personal income. But the difference is less than might have been expected.

(3) An application of the model to quarterly data on personal gross saving is again successful. The forecast for 1962 given in [167] is shown to have been fairly near the mark and a new forecast is given for 1964.

(4) An application of the model to annual data on company saving again gives plausible results. The analysis of information from different sources yields similar structural coefficients except in the case of λ. With the growth of standardized accounting information for companies, I propose to test the model further by applying it to data for different groups of industries. In doing this I intend to reconsider the treatment of taxation I have used so far and to link the results of this financial study to investment requirements obtained from the study of production.

(5) Equilibrium values of saving ratios have been worked out and the combined saving derived from them has been compared with investment requirements appropriate to a 4 per cent rate of growth. It seems probable that the required rate of saving would be met but that personal saving would play a much larger part and company saving a much smaller one than at present in the composition of private saving.

(6) In reaching this conclusion I have ignored all transient effects due to the fact that at present wealth–income ratios do not assume the equilibrium values appropriate to 4 per cent growth. This is all right for a long-run conclusion but must be reconsidered if we are talking about the next few years. The information needed to allow for transient effects is given in the paper.

(7) I hope before long to be able to check some of the conclusions reached in this paper by applying the model to a more extended range of data. I have mentioned one possibility under (4) above; another is to push the annual and quarterly series further back and see how far the coefficients and even the model itself are affected when asked to relate a larger body of information.

APPENDIX I

THE MATHEMATICAL MODEL

The model used in this paper is the same as the one set out in [167] with a small number of generalizations and extensions. In order to make this paper self-contained, I shall now give a mathematical account of the model to supplement the descriptive account given in the body of the paper.

Let us denote observed income, wealth and expenditure by μ, ω and ϵ respectively, and let us divide each of these variables into two components:

the permanent component, denoted by the suffix 1, and the transient component, denoted by the suffix 2. Thus

$$\mu \equiv \mu_1 + \mu_2 \tag{8}$$

and similarly for ω and ϵ.

The permanent and transient components cannot be observed. They are assumed to be weighted averages of the corresponding observed values and the permanent components of the preceding period. Thus

$$\mu_1 = \lambda\mu + (1-\lambda)E^{-1}\mu_1 \tag{9}$$

and

$$\omega_1 = \lambda\omega + (1-\lambda)E^{-1}\omega_1 \tag{10}$$

where the constant, λ, is the same in each equation, and E^{-1} denotes an operator which transforms the variable to which it is applied into its value in the preceding period. More generally, $E^\theta \xi(\tau) \equiv \xi(\tau+\theta)$; meaning that E^θ applied to a variable ξ at time τ transforms this variable into its value at time $\tau + \theta$.

(a) *The basic model.* In the simplest case, I assume that the permanent component of expenditure is a homogeneous linear function of the permanent components of income and wealth. Thus

$$\epsilon_1 = \alpha_1\omega_1 + \beta_1\mu_1 \tag{11}$$

where α_1 and β_1 denote constants. I also assume that the transient component of expenditure is a homogeneous linear function of the transient component of income. Thus

$$\epsilon_2 = \beta_2\mu_2 \tag{12}$$

By adding together (11) and (12) we see that

$$\epsilon = \epsilon_1 + \epsilon_2$$
$$= \alpha_1\omega_1 + \beta_1\mu_1 + \beta_2\mu_2$$
$$= \alpha_1\lambda\omega + \alpha_1(1-\lambda)E^{-1}\omega_1 + \beta_1\lambda\mu + \beta_1(1-\lambda)E^{-1}\mu_1 + \beta_2\mu - \beta_2\lambda\mu - \beta_2(1-\lambda)E^{-1}\mu_1 \tag{13}$$

on substituting for μ_1 and ω_1 from (9) and (10) and using (8). If we multiply the second line of (13) by $(1-\lambda)E^{-1}$, we obtain

$$(1-\lambda)E^{-1}\epsilon = \alpha_1(1-\lambda)E^{-1}\omega_1 + \beta_1(1-\lambda)E^{-1}\mu_1 + \beta_2(1-\lambda)E^{-1}\mu - \beta_2(1-\lambda)E^{-1}\mu_1 \tag{14}$$

and if we compare (14) with the last line of (13), we see that we can write

$$\epsilon = \alpha_1\lambda\omega + [\beta_1\lambda + \beta_2(1-\lambda)]\mu - \beta_2(1-\lambda)E^{-1}\mu + (1-\lambda)E^{-1}\epsilon \tag{15}$$

Further, since observed saving, σ say, is equal to observed income minus observed expenditure, we can see from (15) that

$$\sigma = \mu - \epsilon$$
$$= [1 - \beta_1\lambda - \beta_2(1-\lambda)]\mu - \alpha_1\lambda\omega + \beta_2(1-\lambda)E^{-1}\mu - (1-\lambda)E^{-1}\epsilon \tag{16}$$

In fitting this expression to observations, I shall take as the dependent variable not saving but the ratio of saving to income. Dividing both sides of (16) by μ, we see that

$$\frac{\sigma}{\mu} = [1-\beta_1\lambda-\beta_2(1-\lambda)]-\alpha_1\lambda\frac{\omega}{\mu}+\beta_2(1-\lambda)\frac{E^{-1}\mu}{\mu}-(1-\lambda)\frac{E^{-1}\epsilon}{\mu} \qquad (17)$$

which shows that the saving ratio is a general linear function of the ratio of wealth to income, the ratio of the preceding period's income to income in this period and the ratio of the preceding period's expenditure to income in this period.

The purpose of the manipulations leading to (15), (16) and (17) is to express observables in terms of observables and to clear the unobservable permanent and transient components out of the equations. We see that, according to the model that has been developed, expenditure and saving depend on four variables: observed income and observed wealth in the same period, and observed income and observed expenditure in the preceding period.

The constants in (17) are estimated from time series of σ, μ, ω and ϵ. We can see that these constants are made up of the constants which appeared in (9), (10), (11) and (12). These constants can also be estimated unequivocally from the first set, as we shall see under (f) below.

For practical purposes, a number of extensions of the basic model are needed.

(b) *Non-homogeneity*. Equations (11) and (12) are homogeneous and we may wish to consider the introduction of constant terms in each case, so that the two equations are replaced by

$$\epsilon_1 = \alpha_1\omega_1+\beta_1\mu_1+\gamma_1 \qquad (18)$$

and

$$\epsilon_2 = \beta_2\mu_2+\gamma_2 \qquad (19)$$

where γ_1 and γ_2 denote constants. With these changes, (17) is replaced by

$$\frac{\sigma}{\mu} = [1-\beta_1\lambda-\beta_2(1-\lambda)]-\alpha_1\lambda\frac{\omega}{\mu}+\beta_2(1-\lambda)\frac{E^{-1}\mu}{\mu}-(1-\lambda)\frac{E^{-1}\epsilon}{\mu}-\lambda(\gamma_1+\gamma_2)\frac{1}{\mu} \qquad (20)$$

Equation (20) differs from (17) only in the presence of the final term, $-\lambda(\gamma_1+\gamma_2)/\mu$. The values of α_1, β_1, β_2 and λ can be estimated as before but γ_1 and γ_2 cannot be estimated separately; only their sum, $\gamma_1+\gamma_2=\gamma$ say, can be estimated.

(c) *Exogenous influences*. At various times in recent years the government has taken steps to encourage or discourage spending. We can introduce these influences as follows.

Let ξ denote an index of the relevant government measures and let us define a normal level ξ^* of ξ by the relationship

$$\xi^* = \lambda\xi + (1-\lambda)E^{-1}\xi^* \tag{21}$$

where λ denotes the same constant as in (9) and (10). Now let us add a term in $\xi - \xi^*$ to (12), thus replacing (12) by

$$\epsilon_2 = \beta_2\mu_2 + \delta(\xi - \xi^*) \tag{22}$$

where δ denotes a constant. With this change, (17) is replaced by

$$\frac{\sigma}{\mu} = [1-\beta_1\lambda-\beta_2(1-\lambda)] - \alpha_1\lambda\frac{\omega}{\mu} + \beta_2(1-\lambda)\frac{E^{-1}\mu}{\mu} - (1-\lambda)\frac{E^{-1}\epsilon}{\mu} - \delta(1-\lambda)\frac{\Delta^*\xi}{\mu} \tag{23}$$

where $\Delta^* \equiv (1-E^{-1})$, that is an operator which converts the variable to which it is applied into the excess of this year's value over last year's value.

(*d*) *The subdivision of income.* In the simple homogeneous model no allowance was made for the possibility that wage earners might react very differently from profit earners in response to a change in income. With the statistics at present available we can form time series of different types of income but not of the corresponding divisions of wealth or expenditure. This being so, we can extend the simple model only if we assume that the values of α_1 and λ are the same for different types of income recipient.

Let us suppose that there are two types of income recipient denoted by the suffixes a and b. Equations (8) through (12) can be written down for each of the two types and (17) is replaced by

$$\frac{\sigma}{\mu} = [1-\beta_{1a}\lambda-\beta_{2a}(1-\lambda)]\frac{\mu a}{\mu} + [1-\beta_{1b}\lambda-\beta_{2b}(1-\lambda)]\frac{\mu b}{\mu} - \alpha_1\lambda\frac{\omega}{\mu}$$
$$+ \beta_{2a}(1-\lambda)\frac{E^{-1}\mu a}{\mu} + \beta_{2b}(1-\lambda)\frac{E^{-1}\mu b}{\mu} - (1-\lambda)\frac{E^{-1}\epsilon}{\mu} \tag{24}$$

Here again, the constants α_1, β_{1a}, β_{1b}, $\beta_{,2a}$ β_{2b} and λ can be estimated unequivocally.

(*e*) *Time units.* So far I have not been explicit about time units. I have sometimes used the word 'year' for convenience but evidently the model could be applied to quarters or any other period for which data are available. We must realize, however, that if we change the time unit we shall change the model; except under highly restrictive conditions, we cannot aggregate a quarterly model to give an annual model of the same form. This can be seen from the following example in which the simple model is still further simplified. Suppose that

$$\epsilon = \nu_1\mu + \nu_2 E^{-1}\epsilon \tag{25}$$

in which ν_1 and ν_2 are constants. If this equation relates to quarters we can write it out for four consecutive quarters in the following matrix equation

$$
\begin{bmatrix}
\epsilon \\
E^{-1}\epsilon \\
E^{-2}\epsilon \\
E^{-3}\epsilon
\end{bmatrix}
=
\begin{bmatrix}
\nu_1 & \nu_1\nu_2 & \nu_1\nu_2^2 & \nu_1\nu_2^3 & 0 & 0 & 0 \\
0 & \nu_1 & \nu_1\nu_2 & \nu_1\nu_2^2 & \nu_1\nu_2^3 & 0 & 0 \\
0 & 0 & \nu_1 & \nu_1\nu_2 & \nu_1\nu_2^2 & \nu_1\nu_2^3 & 0 \\
0 & 0 & 0 & \nu_1 & \nu_1\nu_2 & \nu_1\nu_2^2 & \nu_1\nu_2^3
\end{bmatrix}
\begin{bmatrix}
\mu \\
E^{-1}\mu \\
E^{-2}\mu \\
E^{-3}\mu \\
E^{-4}\mu \\
E^{-5}\mu \\
E^{-6}\mu
\end{bmatrix}
+ \nu_2^4
\begin{bmatrix}
E^{-4}\epsilon \\
E^{-5}\epsilon \\
E^{-6}\epsilon \\
E^{-7}\epsilon
\end{bmatrix}
\tag{26}
$$

If we denote by ν_1^* and ν_2^* the constants in an annual model corresponding to ν_1 and ν_2 in the quarterly model, we can see that $\nu_2^* = \nu_2^4$ but that, in general, ν_1^* is not defined in terms of ν_1 and ν_2.

(*f*) *Statistical methods.* Equations of the form of (17), (20), (23) and (24), or a variant embodying two or more extensions of the simple model, have been fitted to annual or quarterly time series by the method of least squares. If we write (17) in the form

$$
\frac{\sigma}{\mu} = b_0 + b_1\frac{\omega}{\mu} + b_2\frac{E^{-1}\mu}{\mu} + b_3\frac{E^{-1}\epsilon}{\mu}
\tag{27}
$$

we can express α_1, β_1, β_2 and λ in terms of b_0, b_1, b_2, and b_3 as follows:

$$
\alpha_1 = -b_1/(1+b_3)
\tag{28}
$$

$$
\beta_1 = (1-b_0-b_2)/(1+b_3)
\tag{29}
$$

$$
\beta_2 = -b_2/b_3
\tag{30}
$$

$$
\lambda = (1+b_3)
\tag{31}
$$

(*g*) *Steady-state ratios.* An important feature of the relationships used in this paper is that saving depends on wealth as well as on income. Since the rate of change of wealth is equal to saving we can calculate the wealth–income ratio and the saving–income ratio which would accompany any steady rate of growth in income.

In deriving these relationships, the first step is to work out the connection between income and its permanent component. By continuous substitution for $E^{-\theta}\mu_1$ in (9), we obtain

$$
\mu_1 = \lambda \sum_{\theta=0}^{\infty} (1-\lambda)^\theta E^{-\theta}\mu
\tag{32}
$$

If μ grows exponentially at a constant rate, ρ, that is if

$$
\mu = \bar{\mu}e^{\rho t}
\tag{33}
$$

where $\bar{\mu}$ denotes the value of μ at time $t=0$, then it follows, by combining (32) and (33), that

$$\frac{\mu_1}{\mu} = \frac{\lambda}{1-(1-\lambda)e^{-\rho}}$$
$$= \zeta \tag{34}$$

say. By the same argument ω_1 and ω are connected in the same way.

If we now consider the model consisting of (8), (9), (10), (18) and (19) and write $\dot{\omega}$ for the rate of change of wealth, we can see that, for steady growth in μ,

$$\dot{\omega} = \sigma$$
$$= \mu - \epsilon$$
$$= [1-\beta_1\zeta-\beta_2(1-\zeta)]\mu-\alpha_1\zeta-\gamma$$
$$= \phi\mu-\psi\omega-\gamma \tag{35}$$

say, where $\gamma \equiv \gamma_1+\gamma_2$. The solution of this differential equation is

$$\omega = \kappa e^{-\psi t}+\frac{\phi\mu}{\psi+\rho}-\frac{\gamma}{\psi} \tag{36}$$

or, as a wealth–income ratio,

$$\frac{\omega}{\mu} = \frac{\kappa e^{-\psi t}}{\mu}+\frac{\phi}{\psi+\rho}-\frac{\gamma}{\psi\mu} \tag{37}$$

where

$$\kappa \equiv \bar{\omega}-\left[\frac{\phi\bar{\mu}}{\psi+\rho}-\frac{\gamma}{\psi}\right] \tag{38}$$

that is the difference between the actual and the equilibrium wealth at time 0 given the rate of growth, ρ, we are considering.

If we differentiate (36) with respect to time, we obtain

$$\dot{\omega} = -\kappa\psi e^{-\psi t}+\frac{\phi\rho\mu}{\psi+\rho} \tag{39}$$

or, as a saving–income ratio,

$$\frac{\sigma}{\mu} = -\frac{\kappa\psi e^{-\psi t}}{\mu}+\frac{\phi\rho}{\psi+\rho} \tag{40}$$

We see from (37) and (40) that, under conditions of steady growth in income, the two ratios, ω/μ and σ/μ, are made up of two parts: a transient part, the first term on the right-hand side of each equation, and a steady-state part, the remaining terms or term. If initially, at time $t=0$, the wealth–income ratio takes its equilibrium value, then $\kappa=0$ and the transient terms disappear.

APPENDIX II

THE STATISTICAL DATA AND RESULTS

The statistical data used in this paper and the results obtained are set out in tables 1 through 5.

TABLE 1   PERSONS: DATA AND RESULTS

	Current values						Values at 1958 prices							Saving ratios			
	Saving	Income	Wages and transfers	Other income	Expenditure	Price index	Saving	Wealth	Income	Wages and transfers	Other income	Expenditure	Change in government measures	Actual	Calculated (1)	Calculated (2)	Calculated (3)
1948	−256	8322	6661	1661	8578	0·6914	−370	52790	12037	9634	2403	12407	0	−3·1	—	—	—
1949	−134	8773	7056	1717	8907	0·7050	−190	52420	12444	10009	2435	12634	0	−1·5	−2·2	−2·2	−1·9
1950	−202	9209	7465	1744	9411	0·7246	−279	52230	12709	10302	2407	12988	0	−2·2	−1·4	−1·5	−1·3
1951	−261	9888	8218	1670	10149	0·7908	−330	51951	12504	10392	2112	12834		−2·6	−2·8	−2·8	−2·8
1952	65	10772	8959	1813	10707	0·8367	78	51621	12875	10708	2167	12797	21·5	0·6	−0·0	0·1	0·1
1953	189	11537	9573	1964	11348	0·8521	222	51699	13540	11235	2305	13318	1·8	1·6	1·4	1·3	1·2
1954	120	12175	10178	1997	12055	0·8688	138	51921	14014	11715	2299	13876	−10·7	1·0	1·3	1·2	1·0
1955	234	13217	11063	2154	12983	0·9005	259	52059	14677	12285	2392	14418	−0·6	1·8	2·6	2·7	2·6
1956	518	14227	11990	2237	13709	0·9421	550	52318	15101	12727	2374	14551	25·8	3·6	4·1	4·2	4·1
1957	547	14991	12606	2385	14444	0·9731	562	52868	15405	12954	2451	14843	2·2	3·7	3·6	3·5	3·4
1958	419	15631	13085	2546	15212	1·0000	419	53430	15631	13085	2546	15212	−8·8	2·7	3·1	3·0	3·0
1959	605	16550	13792	2758	15945	1·0045	603	53849	16476	13730	2746	15873	−31·2	3·7	3·3	3·5	3·5
1960	1062	17752	14697	3055	16690	1·0130	1048	54452	17524	14508	3016	16476	6·7	6·0	6·2	6·4	6·6
1961	1415	18946	15725	3221	17531	1·0425	1357	55500	18174	15084	3090	16816	3·0	7·5	6·6	6·6	6·7
1962	1166	19618	16416	3202	18452	1·0822	1077	56857	18128	15169	2959	17051		5·9	5·8	5·5	5·6

Main Source: *National Income and Expenditure*, 1963.   The value figures are in £ million, the saving ratios in percentages.

TABLE 2 PERSONS (GROSS SAVING): DATA AND RESULTS

Year		Current values			Price index	Saving	Values at 1958 prices			Saving ratios	
		Saving	Income	Expenditure			Wealth	Income	Expenditure	Actual	Calculated
1955	1	131	3289	3158	0·8839	148	49518	3721	3573	4·1	—
	2	170	3381	3211	0·8942	190	49666	3781	3591	5·0	4·7
	3	159	3450	3291	0·9074	175	49856	3802	3627	4·6	5·1
	4	163	3486	3323	0·9162	178	50031	3805	3627	4·7	4·7
1956	1	197	3567	3370	0·9263	213	50209	3851	3638	5·5	5·2
	2	221	3649	3428	0·9449	234	50422	3862	3628	6·1	5·5
	3	254	3690	3436	0·9484	268	50656	3891	3623	6·9	6·1
	4	246	3721	3475	0·9489	259	50924	3921	3662	6·6	6·6
1957	1	258	3771	3513	0·9590	269	51183	3932	3663	6·8	6·3
	2	232	3815	3583	0·9689	239	51452	3937	3698	6·1	6·4
	3	220	3868	3648	0·9804	224	51691	3945	3721	5·7	5·8
	4	240	3940	3700	0·9838	244	51915	4005	3761	6·1	6·0
1958	1	233	3974	3741	0·9934	235	52159	4001	3766	5·9	5·8
	2	199	3967	3768	1·0035	198	52394	3953	3755	5·0	5·1
	3	219	4022	3803	0·9997	219	52592	4023	3804	5·4	5·5
	4	173	4073	3900	1·0033	172	52811	4059	3887	4·2	5·7
1959	1	223	4122	3899	1·0098	221	52983	4082	3861	5·4	4·8
	2	270	4229	3959	1·0003	270	53204	4228	3958	6·4	6·8
	3	266	4253	3987	1·0025	265	53474	4242	3977	6·3	6·7
	4	270	4370	4100	1·0056	268	53739	4345	4077	6·2	7·4
1960	1	292	4415	4123	1·0068	290	54007	4385	4095	6·6	7·0
	2	358	4527	4169	1·0144	353	54297	4463	4110	7·9	7·7
	3	424	4605	4181	1·0123	419	54650	4549	4130	9·2	8·9
	4	439	4656	4217	1·0184	431	55069	4572	4141	9·4	9·4
1961	1	468	4780	4312	1·0291	455	55500	4645	4190	9·8	9·9
	2	523	4863	4340	1·0338	506	55955	4704	4198	10·8	10·1
	3	437	4870	4433	1·0490	417	56461	4643	4226	9·0	9·9
	4	474	4920	4446	1·0581	448	56878	4650	4202	9·6	8·9
1962	1	448	4963	4515	1·0740	417	57326	4621	4204	9·0	9·0
	2	368	4991	4623	1·0865	339	57743	4594	4255	7·4	8·3
	3	430	5058	4628	1·0828	397	58082	4671	4274	8·5	7·9
	4	436	5122	4686	1·0852	402	58479	4720	4318	8·5	8·5
1963	1	467	5144	4677	1·0925	427	58881	4708	4281	9·1	8·2
	2	434	5242	4808	1·0960	396	59308	4783	4387	8·3	9·1
	3	470	5345	4875	1·0930	430	59704	4890	4460	8·8	8·9
	4					448	60134	4944	4496		9·1
1964	1					447	60582	4960	4513		9·0
	2					457	61029	5005	4548		9·1
	3					466	61486	5050	4584		9·2
	4					477	61952	5100	4623		9·4

Main Source: *Economic Trends*, No. 120, October, 1963. The value figures are in £ million, the saving ratios in percentages. The forecast values at 1958 prices are shown below the bars.

TABLE 3   ALL COMPANIES: DATA AND RESULTS

	Current values			Price index	Values at 1958 prices				Saving ratios	
	Saving	Income	Expenditure		Saving	Wealth	Income	Expenditure	Actual	Calculated
1948	783	1295	512	0·6914	1133	14281	1872	741	60·5	—
1949	671	1165	494	0·7050	952	15414	1653	701	57·6	58·9
1950	1111	1618	507	0·7246	1533	16366	2233	700	68·7	67·6
1951	1436	1983	547	0·7908	1816	17899	2508	692	72·4	71·0
1952	818	1351	533	0·8367	978	19715	1615	637	60·6	61·0
1953	906	1482	576	0·8521	1063	20693	1739	676	61·1	62·3
1954	1225	1856	631	0·8688	1410	21756	2136	726	66·0	65·9
1955	1288	1976	688	0·9005	1430	23166	2194	764	65·2	65·2
1956	1373	2084	711	0·9421	1457	24596	2212	755	65·9	63·6
1957	1325	2079	754	0·9731	1362	26053	2137	775	63·7	62·6
1958	1215	2008	793	1·0000	1215	27415	2008	793	60·5	59·2
1959	1376	2284	908	1·0045	1370	28630	2274	904	60·2	61·5
1960	1799	2931	1132	1·0130	1776	30000	2894	1118	61·4	64·5
1961	1344	2588	1244	1·0425	1289	31776	2482	1193	51·9	53·9
1962	1200	2469	1269	1·0822	1109	33065	2282	1173	48·6	46·5

Main Source: *National Income and Expenditure*, 1963.   The value figures are in £ million, the saving ratios in percentages.

TABLE 4 QUOTED COMPANIES IN MANUFACTURING AND DISTRIBUTION: DATA AND RESULTS

	Current values				Price index	Values at 1958 prices				Saving ratios	
	Saving	Wealth	Income	Expenditure		Saving	Wealth	Income	Expenditure	Actual	Calculated
1949	270	4133	422	152	0·7050	383	5862	599	216	64·0	—
1950	344	4472	510	166	0·7246	475	6172	704	229	67·4	68·3
1951	391	4956	560	169	0·7908	494	6267	708	214	68·4	68·2
1952	287	5439	463	176	0·8367	343	6500	553	210	62·0	63·7
1953	336	5822	540	204	0·8521	394	6832	634	240	62·2	62·6
1954	432	6208	676	244	0·8688	497	7145	778	281	63·9	64·0
1955	493	6701	760	267	0·9005	547	7441	844	297	64·9	64·5
1956	465	7368	740	275	0·9421	494	7821	786	292	62·8	62·0
1957	448	7993	750	302	0·9731	461	8214	771	310	59·7	59·3
1958	483	8633	795	312	1·0000	483	8633	795	312	60·8	57·9
1959	574	9226	992	418	1·0045	571	9185	987	416	57·9	58·8
1960	660	10009	1058	458	1·0130	592	9881	1044	452	56·7	59·1

Main Source: *Economic Trends*, No. 102, April 1962. The value figures are in £ million, the saving ratios in percentages.

TABLE 5   EQUILIBRIUM SAVING – INCOME AND WEALTH –
INCOME RATIOS FOR DIFFERENT GROWTH RATES OF INCOME

100ρ	Persons		All companies		Quoted companies	
	$100\sigma/\mu$	ω/μ	$100\sigma/\mu$	ω/μ	$100\sigma/\mu$	ω/μ
0	0	3·8	0	15·3	0	22·0
1	3·4	3·4	13·4	13·4	17·2	17·2
2	6·1	3·1	23·8	11·9	28·3	14·1
3	8·4	2·8	32·1	10·7	35·9	12·0
4	10·2	2·6	38·7	9·7	41·6	10·4
5	11·8	2·4	44·2	8·8	45·9	9·2
6	13·2	2·2	48·7	8·1	49·4	8·2
7	14·5	2·1	52·5	7·5	52·1	7·5
8	15·6	2·0	55·8	7·0	54·4	6·8
9	16·5	1·8	58·6	6·5	56·4	6·3
10	17·4	1·7	61·0	6·1	58·0	5·8
∞	43·9	0	90·1	0	74·9	0

MULTIPLE CLASSIFICATIONS IN SOCIAL ACCOUNTING

1. INTRODUCTION

A complete system of social accounts must be able to handle transactors in all their aspects: as producers, consumers and accumulators. To reduce the number and variety of transactors to manageable dimensions it is necessary to classify them, but experience shows that it is impossible to find a single classification which will be equally suitable for each aspect. This presents a real problem both to the statistician and to the analyst, a problem which seems to admit of only three solutions, as follows.

(a) *The limited solution.* We can remove altogether, or reduce to a minimum, such as public and private, the transactor classification in the accounting system. Subsidiary detail can then be set out in supplementary tables in which a variety of classifications can be used. Thus production can be classified by industry, consumption by commodities, lending and borrowing by financial sectors, and so on. The totals in the supplementary tables agree with the corresponding entries in the accounting system but one does not ask embarrassing questions about the connection between individual items in the supplementary tables. If we follow this line of approach we cannot, for example, trace the connection between the consumption of fruit and vegetables and the production of the food processing industries, nor discover through what channels a new industry is financed. This method, however, though limited is a practical one and is adopted in the international standard systems of national accounts [118, 181].

(b) *The solution of Procrustes.* We can adopt a single classification of transactors and apply it everywhere whether it is suitable or not. Where a statistical system is centred on production statistics and where inter-industry relationships are considered of particular interest, it is very tempting to follow this method. But it has serious disadvantages. For example, products which one would like to classify together in the analysis of consumption do not necessarily come from a single industry. Indeed, with the treatment of trade and transport as margin activities, as in input–output, they almost never do.

(c) *The proper solution.* We can choose as many classifications as we think would be useful and put them all into the accounting system. Thus we might have *r* rows and columns for industries, *s* for commodities, *t* for financial sectors, and so on. The resulting system will be very large and can be filled in in various ways, connecting and generalizing simple systems in various degrees. At the intersection of two classification systems, say the rows of *r* and the columns of *s*, we shall get a connection between the *r* and the *s* classification. If this connection is reduced to a set of constants, or functions of exogenous variables, we shall have a classification converter from which we shall be able to calculate requirements in the *r* classification from given levels expressed in the *s* classification.

In this paper I shall explore the proper solution and try to show: (*a*) that it is theoretically enlightening; (*b*) that it is extremely flexible and has many practical advantages, in particular that it enables us to deal with problems on their own ground and to be quite open about the complexities of the real world that lead us to want several classifications; and (*c*) that it can be put into practice with the kind of information which is typically available at least in countries with advanced systems of economic statistics.

2. CRITERIA OF CLASSIFICATION FOR A PROPER SOLUTION

I believe that the problem of choosing suitable criteria of classification can best be viewed as a question of supply and demand. The supply side is determined by the practicability of providing information compiled on different conceptual bases, and this is limited not only by the difficulties and cost of collection but also by the difficulties of rearranging the crude data in the course of processing them. The demand side is determined by the model of the economic process which we should like to construct. In the past, economic statisticians who collect and process data and economic theorists who design models of the economic process have tended to live in separate worlds. Fortunately this state of affairs is breaking down, and the sooner it breaks down altogether, the better. The statistician should be able to understand why certain distinctions are asked for by the theorist and should be ready with a constructive alternative if he cannot meet demands *au pied de la lettre*. Equally, the theorist should be able to understand why some distinctions are very difficult to make in practice and should be ready with a constructive alternative if his demands cannot be met exactly.

Everyone will agree that a social accounting system must preserve the distinctions between production, consumption and accumulation. A transactor may enter into any of these activities, and to preserve the distinctions for him is a matter of drawing up a suitable set of accounts. But we must then face the problem of combining these accounts: how can this best be done?

If we start with production there seems little doubt that the model-builders will press for a definition of a branch of production which is expressed in terms of a set of commodities. The reason is that they would like the connections between branches of production to be as far as possible technological because then they would be relatively stable, certainly more stable than if they were modified by administrative and institutional factors. This line of thought usually leads to the choice of the establishment as the basic unit in the collection of statistics of input and output. This means that information about the purchases, sales and stocks of various products is collected from establishments or workplaces. Each establishment is then assigned to a trade in accordance with its characteristic, or principal, products and the trades are grouped into industries. An attempt may then be made to define the branches of production in terms of principal–product groups and to move subsidiary products and by-products out of the trades or industries in which they are recorded into the trades or industries of which they are principal products. Ways of doing this have been worked out by input–output analysts, as explained in [151]. The final stage is to try to replace purchases by absorptions so that input–output relationships are not distorted by purchases which differ from absorptions in current production. To the extent that this programme of modification can be carried out, industries come to be defined largely in technological terms, though in practice their principal product is never completely homogeneous.

If we now pass to the activity of consumption, including all redistributions of income, the position is quite different. For here we are usually interested in connections which are financial rather than technological. Thus we might begin by dividing consumers into private and public and then divide each of these categories along institutional lines. But we should then have to face the fact that a consumers' classification of products is not the same as a producers' classification. When we analyse the demand of private consumers we begin by grouping commodities on the lines of a shopping list; when we analyse the demand of public consumers we think of it in terms of purposes, such as education, public health or defence. But each of these commodities or purposes typically spans a number of product categories defined in technological terms; and this is true even apart from the treatment of trade and transport as margin activities, which is inevitable in input–output analysis. Accordingly, it does not seem desirable to set up consumers' classifications of products in the same terms as we have used in classifying production. It seems better to set them up in their own terms and then convert them into the relevant producers' categories.

A similar situation holds on the income side of consumption accounts, since here the technologically defined industries must pay income into financially defined sectors. In the case of wages this is a simple matter since all wage-earners are in the household sector. But with profits it is much more difficult because each sector may be represented in each industry. One way round this difficulty is to have an appropriation account for business as a whole, which then pays out income to each financially defined sector. This is not a solution, it is only a way of cutting the Gordian knot: it enables one to study financial flows from industries to sectors, but not flows between industries defined either financially or in any other way.

When we come to accumulation, the story we have been through with the current accounts largely repeats itself. If we are interested in production functions we must concern ourselves with capital goods produced by different industries, technologically defined, which must be put at the disposal of other industries defined in the same way. This means that we need a set of connections between the production and accumulation accounts of technologically defined industries. But if we are interested in finance, in the sources and uses of saving, in the way in which producers are related financially to households, government and financial intermediaries, we need a set of institutional capital accounts.

Thus I am arguing for a unified system of social accounting in which the classifications adopted at different points are appropriate to their subject matter. Speaking as an econometrician, I think that national accounts plus supporting tables are unsatisfactory for model-building because one needs to know the accounting restrictions of the system in much greater detail by sector. Equally I am opposed to attempts to work with a single system of classification which, in so far as it is appropriate to anything, usually turns out to be appropriate almost exclusively to input–output analysis. This I have called the solution of Procrustes. Worst of all perhaps is the state of mind in which the importance of theoretical distinctions is belittled, with the result that attempts are made at unnecessary compromises and data are produced that do not fit any part of the conceptual system they are intended to clothe.

3. A SIMPLE EXAMPLE: INDUSTRIES AND COMMODITIES

In section 6 below I shall give a numerical example of a social accounting system which embodies several different kinds of classification, but before we come to this it may be helpful to give some simple examples which concentrate on one or two points at a time. Let us consider to begin with three sets of accounts: the first relating to industries; the second relating to commodities; and the third relating to final consumers. Let the typical industry be denoted by the suffix r and let the typical commodity be denoted by the suffix s. Then the flow matrix of the system can be written in the following partitioned form:

$$\begin{bmatrix} Z_{rr} & Z_{rs} & e_r \\ \hline Z_{sr} & Z_{ss} & e_s \\ \hline g'_r & g'_s & 0 \end{bmatrix}$$

where the Z's are matrices of product flows, the e's are vectors of final demands and the g's are vectors of primary inputs (the primes indicate that g'_r and g'_s are row vectors). Thus Z_{rr} denotes the inter-industry flows, Z_{rs} denotes the amounts of industry outputs embodied in different commodities, and so on. If we suppose that transactions take place only in commodities, then the above matrix takes the form

$$\begin{bmatrix} 0 & \vdots & Z_{rs} & \vdots & 0 \\ \cdots & \vdots & \cdots & \vdots & \cdots \\ Z_{sr} & \vdots & 0 & \vdots & e_s \\ \cdots & \vdots & \cdots & \vdots & \cdots \\ g'_r & \vdots & 0 & \vdots & 0 \end{bmatrix}$$

In this matrix all outputs are converted into commodities, the elements of Z_{rs}, and all commodities are either used by industries as intermediate product, the elements of Z_{sr}, or go to final consumers, the elements of e_s. Only industries require primary inputs, the elements of g_r.

Given the industry outputs, q_r say, and the commodity outputs, q_s say, we can form the coefficient matrices $A_{rs} = Z_{rs}\hat{q}_s^{-1}$ and $A_{sr} = Z_{sr}\hat{q}_r^{-1}$, where $\hat{q}_s^{-1}$ and $\hat{q}_r^{-1}$ are diagonal matrices constructed from the reciprocals of the elements of q_s and q_r respectively. Thus

$$q_r = A_{rs}q_s \tag{1}$$

and

$$q_s = A_{sr}q_r + e_s$$
$$= A_{sr}A_{rs}q_s + e_s$$
$$= (I - A_{sr}A_{rs})^{-1}e_s \tag{2}$$

From (1) and (2) it follows that

$$q_r = A_{rs}(I - A_{sr}A_{rs})^{-1}e_s$$
$$= (I - A_{rs}A_{sr})^{-1}A_{rs}e_s \tag{3}$$

which gives q_r in terms of e_s with a matrix multiplier of order equal to the number of commodities or the number of industries.

In the numerical example in section 6 below we do not go as far as this. In that example only final consumers buy in terms of the commodity classification; producers buy and sell in terms of the industrial classification. Thus the flow matrix takes the form

$$\begin{bmatrix} Z_{rr} & \vdots & Z_{rs} & \vdots & 0 \\ \cdots & \vdots & \cdots & \vdots & \cdots \\ 0 & \vdots & 0 & \vdots & e_s \\ \cdots & \vdots & \cdots & \vdots & \cdots \\ g'_r & \vdots & 0 & \vdots & 0 \end{bmatrix}$$

In this case Z_{rr} is the usual matrix of inter-industry flows and Z_{rs} is the consumers' classification converter. Thus a row of Z_{rs} shows how the products of an industry contribute to different commodities in the consumers' classification and a column of Z_{rs} shows how one item in the consumers' classification is made up of the outputs of different industries. Final demands in terms of industry outputs appear as the elements of $e_r = Z_{rs}i$, where i is the unit vector, that is the row sums of Z_{rs}. Demand analysis can be used to project the elements of e_s; a coefficient matrix, A_{rs} say, possibly with elements that change over time, can be used to obtain e_r; and the industry outputs, q_r, can be estimated by the usual matrix-multiplier operation. Evidently this method can be applied with any number of consumers' classifications.

There are several advantages in using a number of classifications in this way. First, demand analysis can be applied to consumers' categories: we do not have to attempt a demand analysis for mechanical engineering or for chemicals, still less for trade or transport margins, but can concentrate on the different consumers' categories into which these products enter. The precise way in which consumers' demands are met can then be treated as a separate problem: how the demand for heat and light is met from coal, gas, electricity or oil. Second, by introducing two classifications of commodities we have some latitude in the treatment of direct labour, complementary imports and indirect taxes. Thus the indirect taxes on whisky and gin can best be taken out in the classification converter which will now extend to the bottom row and middle column of the matrix. The reason is that the distilling industry also produces large quantities of industrial alcohol which is not subject to tax. On the other hand, taxes on the petrol used by public transport are best taken out at the inter-industry stage, since this tax will have to be paid whoever pays for the public transport.

4. A SECOND EXAMPLE: A CLASSIFICATION CONVERTER FOR FINANCIAL FLOWS

In the preceding section we have been concerned with the conversion of real flows from one classification to another. In a similar way we can imagine a classification converter for financial flows. Consider the major sectors of government: central and local and, within local, urban and rural. Consider, further, the building of roads. Each sector of government may have a direct responsibility to build and maintain certain roads. And in addition there may be certain fixed transfers from one type of authority to another: the central government might make a grant of such and such a proportion of the expenditure on roads by local authorities; urban authorities, which are usually richer than rural authorities, might make a grant of such and such a proportion of the rural expenditure on roads in their area. If all these proportions were known, as in practice they are, we could convert the actual expenditures on roads by the different types of authority into the sums of money which each type would have to raise to meet not only its own expenditure on roads but also its financial commitments consequent on the expenditure of other authorities. This means that government purposes turn up twice in our classification scheme. The first conversion, treated implicitly in the preceding section, connects government expenditures on education, health, roads, etc., with the industries which are stimulated by these expenditures. The second conversion connects the same set of government expenditures with the financial demands they impose on different sectors of goverment.

This kind of converter may be important in the study of financial relationships where there are transfer commitments between different financial sectors. It will not be illustrated in the example in section 6 because the model for which that example provides an accounting framework is more concerned with real flows than with the details of their financial implications.

5. REAL-FINANCIAL CONVERTERS

In the preceding sections I have given examples of classification converters on the real and on the financial side of the economy. Since the two sides interact there must naturally be real-financial and financial-real converters. An obvious illustration of the former is the conversion of the income paid out by industries into income received by financial sectors. In the example given in section 6 these sectors are broadly defined as corporations, households, central government and so on, but in an accounting system directed in particular to detailed financial relationships they might include industries defined in financial terms, that is as enterprises, concerns or even wider financial groupings, which might have very little in common with industries defined in technological terms. An illustration of a financial-real converter is the conversion of capital finance by sectors into the net investment in fixed assets and stocks in different industries.

6. A NUMERICAL EXAMPLE: THE ACCOUNTING FRAMEWORK OF THE CAMBRIDGE GROWTH MODEL

This model is an econometric model of the British economy with particular emphasis on real flows. Its object is to provide a means of examining the implications of alternative rates of growth in total consumption. It is in its early stages and has been outlined in [152]. From that brief account it can be seen to rest on three main features: (i) an accounting framework; (ii) a set of econometric relationships covering demand, input–output, production functions, asset requirements, etc.; and (iii) a computing programme. As will be seen, there are only a modest number of categories in each classification. The reason is that we wished to construct within a reasonable time a complete, computable model. If we had tried to make it larger at the outset we should have increased our difficulties over data and should also have gone beyond the capacity of the high-speed memory of the Cambridge computer, EDSAC. As it is we are able to reach a solution in a two-cycle feed-back process in approximately four minutes' computing time. It is thus practicable to investigate many alternative hypotheses relating to the structure of demand, the level of exports, the industrial distribution of labour, etc.

The purpose of this section is to explain the framework of our model, which illustrates the use of multiple classifications; not to explain the model itself. This framework consists of a social accounting system with 87 balancing accounts. These are listed in order in the appendix. The whole matrix of transactions, T say, can conveniently be partitioned into 49 sub-matrices corresponding to the seven major categories of the system. In practice 27 of these submatrices will be empty and the whole system can be presented, for purposes of description, as follows:

$$T = \begin{bmatrix} T_{11} & T_{12} & T_{13} & 0 & T_{15} & 0 & T_{17} \\ 0 & 0 & 0 & T_{24} & 0 & 0 & 0 \\ 0 & 0 & 0 & T_{34} & 0 & 0 & 0 \\ T_{41} & T_{42} & T_{43} & T_{44} & 0 & 0 & 0 \\ T_{51} & 0 & 0 & 0 & 0 & T_{56} & 0 \\ 0 & 0 & 0 & T_{64} & 0 & T_{66} & T_{67} \\ T_{71} & T_{72} & T_{73} & T_{74} & 0 & T_{76} & T_{77} \end{bmatrix} \qquad (4)$$

The actual appearance of T is shown in the provisional table for 1959 at the end of this paper.

The seven major categories are as follows:

(*a*) *Industries' current accounts* (*production accounts*), of which there are 18. The first 18 rows and columns thus show the incomings and outgoings of different branches of production.

(*b*) *Consumers' goods*, of which there are 13. Rows 19 through 31 each contain one entry, equal to the expenditure on a particular class of consumers' good (or service). In columns 19 through 31 each item is subdivided into demands on industries, direct labour, net indirect taxes and direct imports of non-competitive goods such as tropical fruit, wine, etc.

(*c*) *Government purposes*, of which there are 15. Rows and columns 32 through 46 do for government expenditure on current goods and services (classified by purpose) what rows 19 through 31 and the corresponding columns do for consumers' goods.

(*d*) *Institutions' current accounts* (*income accounts*), of which there are 9. Rows and columns 49 through 55 relate to the sectors of the *Blue Book* [176], that is to say companies, persons, public corporations, central government trading, central government non-trading, local authorities trading and local authorities non-trading. Rows and columns 47 and 48 relate respectively to the distribution of non-labour income other than the rent of dwellings and to the distribution of income from dwellings. In these two accounts income is received from its various sources, that is industries, and allocated in total over destinations, that is sectors. Obviously it would be much better to make a to-whom-from-whom classification, but this is difficult in respect of the first category of income because the *Blue Book* does not give enough detail.

(*e*) *Industries' capital accounts*, of which there are again 18. Rows 56 through 73 show the finance, in terms of depreciation and funds for net investment, of the gross investment of each industry, which appears in one of the columns 56 through 73.

(*f*) *Institutions' capital accounts*, of which there are 11. The last seven of these, namely rows and columns 78 through 84, relate again to the *Blue Book* sectors. The first four, namely 74 through 77, relate to the sources and destinations of the finance of different types of investment. Rows 74 through 76 show the institutional sources of finance for net fixed investment in industry, net investment in dwellings and net investment in stocks and the destination of this finance in industries. Row 77 shows the net acquisition of financial claims by sectors together with the net acquisition by the rest of the world and the residual error. The sum of all these acquisitions (including the residual error) is, of course, zero.

(*g*) *The rest of the world.* For completeness, the transactions which the rest of the world has with Britain must be fully accounted for, though of course those between different parts of the rest of the world disappear by consolidation. The relevant transactions are shown in rows and columns 85 through 87. The first of these contains only imports and exports, and the balance on this account (the rest of the world's export surplus to Britain) is transferred to the second account. This second account contains, in addition, factor-income payments, direct taxes and other current transfers, and its balance (the rest of the world's surplus on current account with Britain) is transferred to the third account which, otherwise, contains only the net acquisition of claims on, and capital transfers to, Britain.

7. CONCLUSIONS

(1) I have argued in favour of adopting multiple classifications in social accounting, rejecting both the method of supplementary tables, which cannot show in detail how the whole system fits together, and the method of forcing all elements in the system into a common classification, which cannot be suitable in all cases.

(2) My arguments are based on attempts to use economic statistics in the description and analysis of the economic process. In doing this one finds the need to trace detailed connections and also the possibility of making better use of existing statistics than can be achieved by a single classification system. I am not impressed either by purely theoretical arguments which do not concern themselves with the problems of data collection and processing nor with purely practical arguments which do not concern themselves with theoretically desirable distinctions.

(3) I have shown how different classifications can be related by classification converters. My main example has been the conversion of a consumers' classification of commodities into a producers' one. It is implicit in my argument that one should carry out demand analysis with the most appropriate categories and then discover the production implications of one's results; of course if one starts with a given set of demands already set out in a producers' classification this problem does not arise.

(4) I have emphasized the distinction between real and financial classifications because it seems to run through all the analyses of the economic process that we try to make. I cannot imagine a satisfactory input–output table based on a financial grouping of industries and following the usual principles of financial accounting. Equally I cannot imagine anyone wanting to study financial capital relationships between industries grouped on a technological basis. In fact I find it difficult to see why anyone should want a detailed financial classification of industries at all; but this is probably due to the fact that I am more interested in the details of the real side of the economy than I am in the details of its financial side.

(5) In any case I am very much in favour of production statistics being collected in such a way that they illuminate the real side of the economy which can be stereotyped in terms of input–output analysis. I think that if they are not collected in this way, attempts to use the results for a real analysis will generally prove disappointing.

(6) It is precisely at the intersections of real and financial classifications that it seems particularly difficult, with existing statistics, to avoid the use of dummy accounts which replace a to-whom-from-whom classification of real into financial (or financial into real) with the marginal totals of such a classification. I am not sure that anything of substance is lost by this device. If further details are needed they can be provided by combining hypothesis with observation. With the statistics we possess at present, if both classifications are detailed, hypothesis will play a relatively important role and the results may not be of much value.

(7) All this goes to show that it is very difficult to separate facts and theories. Those who insist on being guided by facts alone may find that they get very little guidance and those who insist on the distinctions of pure theory may confine themselves to a world which most of their fellow men consider uninteresting.

FURTHER DETAILS OF THE ACCOUNTING FRAMEWORK OF THE
CAMBRIDGE GROWTH MODEL

1. A LIST OF THE ACCOUNTS

The main headings, preceded by *a* to *g* in round brackets, correspond to the
classes described in section 6 above.

(*a*) *Industries, current accounts:*

1. Agriculture, forestry and fishing.
2. Mining and quarrying; bricks, pottery, glass, cement, etc.
3. Food, drink and tobacco.
4. Chemicals and allied trades.
5. Metal manufacture.
6. Engineering and electrical goods; metal goods n.e.s.
7. Shipbuilding and marine engineering; aircraft; railway vehicles.
8. Road vehicles.
9. Textiles.
10. Leather, leather goods and fur; clothing and footwear.
11. Timber, furniture, etc.
12. Paper, printing and publishing; other manufacturing industries.
13. Construction.
14. Gas, electricity and water.
15. Transport and communication.
16. Distributive trades.
17. Services.
18. Ownership of dwellings.

(*b*) *Consumers' goods:*

19. Unprocessed food: meat and bacon; fish; dairy products; fruit;
    potatoes and vegetables.
20. Processed food; bread and cereals; oils and fats; sugar, preserves
    and confectionery; beverages; other manufactured food; food
    eaten in meals away from home.
21. Alcoholic drink and tobacco.
22. Housing.
23. Fuel and light.
24. Clothing and footwear.
25. Energy-using household durable goods.
26. Other household durable goods.
27. Private vehicles and their running expenses.

28. Public transport.
29. Other goods.
30. Other services.
31. Income in kind n.e.s.

(c) *Government purposes:*

32. Finance and tax collection.
33. Police and justice.
34. Overseas services.
35. Military and civil defence.
36. Housing.
37. Education and child care.
38. Health services.
39. National insurance, pensions, etc.
40. Agriculture and food.
41. Fuel and power industries.
42. Transport and communication.
43. Other industry and trade.
44. Employment services.
45. Roads and public lighting.
46. Other.

(d) *Institutions, current accounts:*

47. Distribution of profits, interest, etc.
48. Distribution of rent on dwellings.
49. Companies.
50. Persons.
51. Public corporations.
52. Central government trading.
53. Central government non-trading.
54. Local authorities trading.
55. Local authorities non-trading.

(e) *Industries, capital accounts:*

56. Agriculture, forestry and fishing.
57. Mining and quarrying; bricks, pottery, glass, cement, etc.
58. Food, drink and tobacco.
59. Chemicals and allied trades.
60. Metal manufacture.
61. Engineering and electrical goods; metal goods n.e.s.
62. Shipbuilding and marine engineering; aircraft; railway vehicles.
63. Road vehicles.

64. Textiles.
65. Leather, leather goods and fur; clothing and footwear.
66. Timber furniture, etc.
67. Paper, printing and publishing; other manufacturing industries.
68. Construction.
69. Gas, electricity and water.
70. Transport and communication.
71. Distributive trades.
72. Services.
73. Ownership of dwellings.

(*f*) *Institutions, capital accounts:*

74. Net investment in industry.
75. Net investment in dwellings.
76. Net investment in stocks.
77. Net acquisition of claims.
78. Companies.
79. Persons.
80. Public corporations.
81. Central government trading.
82. Central government non-trading.
83. Local authorities trading.
84. Local authorities non-trading.

(*g*) *The rest of the world:*

85. Production.
86. Consumption.
87. Accumulation.

2. THE ENTRIES IN THE ACCOUNTS

The entries in the accounts can conveniently be described by taking in order the submatrices of the matrix of transactions (T in equation (4) of section 6 of the main paper). The elements of these submatrices are denoted, where necessary, by their matrix suffix in T; for example, (13.65) is the entry in row 13 and column 65, that is the gross investment in buildings and works by the leather, clothing and footwear group of industries.

T_{11}. This is the inter-industry submatrix; its entries are all intermediate product flows. In principle they relate to absorptions by principal product groups. Investments in stocks, which appear diagonally in T_{15}, are estimated on this basis.

It must not be supposed that it is possible in Britain to construct a complete inter-industry matrix with a lag of only eighteen months. The entries in T_{11} are constructed from a matrix for 1954 by an approximate method based on the assumption that changes in input–output coefficients through time are the result of three factors: (i) price changes; (ii) changes in the absorption of particular products which apply to all users with a common factor of proportionality; and (iii) changes in the degree of fabrication.

If A_0 is an initial coefficient matrix and if p is a price vector in which current prices are related to initial prices, then the initial matrix converted to current values, A^* say, is given by

$$A^* = \hat{p} A_0 \hat{p}^{-1} \tag{5}$$

that is, by a similarity transformation of A_0. If A^* differs from the current coefficient matrix, A say, because all the users of a given product have changed their absorption per unit of output by a common proportion, then A is equal to A^* premultiplied by a diagonal matrix. If, on the other hand, A^* differs from A because the degree of fabrication of a given set of inputs has changed by a certain proportion in each industry, then A is equal to A^* postmultiplied by a diagonal matrix. If both these factors have been at work, then

$$A = \hat{r} A^* \hat{s}$$
$$= \hat{r}\hat{p} A_0 \hat{p}^{-1} \hat{s} \tag{6}$$

where $\hat{r}$ and $\hat{s}$ are diagonal matrices constructed from vectors r and s.

Consider now an intermediate output vector, u say, and an intermediate input vector, v say, which are the marginal totals of the submatrix of inter-industry flows. Then, for the current period we have

$$Aq = u \tag{7}$$

and

$$\hat{q} A' i = v \tag{8}$$

By hypothesis we do not know the elements of A, the coefficient matrix for the current period. We should like to estimate it from A^* and to do this we might proceed as follows. If q is the current output vector, let us make an initial estimate of u, u_0 say, by premultiplying q by A^*. Thus

$$A^* q = u_0 \tag{9}$$

In general $u_0 \neq u$ and so the next step is to force an equality by an appropriate multiplication of the rows of A^*. Thus

$$\hat{u} \hat{u}_0^{-1} A^* q = u \tag{10}$$

The inter-industry matrix now satisfies the row conditions but not the column conditions. These can be forced by substituting for A from (10)

17

into (8) and by making an appropriate multiplication of the columns of A. Thus

$$\hat{q} A^{*\prime} \hat{u} \hat{u}_0^{-1} i = v_0 \tag{11}$$

and

$$\hat{q} \hat{v} \hat{v}_0^{-1} A^{*\prime} \hat{u} \hat{u}_0^{-1} \imath = v \tag{12}$$

This step will unbalance the rows and the following step will unbalance the columns, but it seems intuitively plausible that a repetition of these steps will bring the matrix more and more into balance. I have not been able to prove this result though it is stated to hold by Deming [47, pp. 115–17]. After $(n+1)$ rounds we shall obtain

$$(\hat{u}^{n+1} \hat{u}_0^{-1} \dots \hat{u}_n^{-1} A^* \hat{v}^{n+1} \hat{v}_0^{-1} \dots \hat{v}_n^{-1}) q = u_{n+1} \tag{13}$$

If in fact this process is to converge, then $u_n \to u$ and $v_n \to v$ as n increases. In this case we may take the term in round brackets as an estimate of A. With this value of A, the marginal conditions (7) and (8) will be satisfied.

The value of A given by (13) can be written in the form

$$A = \hat{r}^* A^* \hat{s}^* \tag{14}$$

Thus $r^* = \lambda r$ and $s^* = \lambda^{-1} s$, where λ is a constant. The value of A in (14) is therefore the same as the value of A in (6).

The elements of T_{11} are obtained from (14) using a coefficient matrix for 1954 as a starting point. The resulting T_{11} will not be correct because the proportionalities on which it is based are only approximations. If, for example, electricity is replacing coal as a source of power, it is likely that this will happen to a greater extent in new and growing industries than in old and stagnant industries. Nevertheless A is an improvement on A^* as an estimate of the current coefficient matrix. Furthermore, the calculation of r^* and s^* provides a basis for calculating future values of A. This basis is strengthened if we have two or more past values of A^* and can study the development of the elements of r^* and s^*. The appropriateness of the basic assumption can to some extent be checked by working forwards and backwards between past values of A^*. If the assumption is correct then, apart from a scalar multiplier, the two premultiplying matrices should be reciprocal, and so should be the two postmultiplying ones.

T_{12}. This is the classification converter for private consumers; its entries show the demands on different industries resulting from different categories of private consumers' expenditure. Thus, for example, (2.23), (4.23), (14.23), (15.23), and (16.23) show respectively the demand on mining, chemicals, gas and electricity, transport and distribution associated with consumers' expenditure on fuel and light. As far as possible these entries are based on the *Blue Book*, but in some cases it is necessary to obtain a subdivision from other sources and it is always necessary to consult other sources for trade and transport margins. Direct labour is not included in this submatrix at all but appears in (50.30). Only some of the indirect taxes (net) which are levied on consumers' goods are included in this submatrix: they are embodied in the sales of the industries concerned and are then debited to them in T_{41}. Those taxes which are excluded from T_{12} appear in T_{42}. Thus petrol duties on road passenger transport are included in (15.28) and, as a cost of transport and communication, in (53.15); in this way the basis for valuing transport services in row 15 is kept as uniform as possible. On the other hand customs and excise duties on drink and tobacco are not included in (3.21) but are shown separately in (53.21); this helps to keep the basis for valuing the output of the distilling industry in row 3 as uniform as possible, since potable alcohol is highly taxed whereas industrial alcohol is not. Finally, certain imports which appear in consumers' expenditure do not appear in T_{12}. These are complementary imports required only by consumers, such as tropical fruit and foreign wine. Competitive imports for final consumers appear positively in T_{12} and negatively in T_{17}; Italian knitwear, for example, is included in (10.24) and subtracted from British exports in (10.85).

T_{13}. This is the classification converter for public authorities' consumption; its entries show the demands on different industries resulting from different categories of expenditure on goods and services by public authorities. The estimates are based on the *Blue Book* and on government accounting publications. Again, direct labour is not included in this submatrix but is shown in row 50. Certain complementary imports are shown in row 85.

T_{15}. The entries in this submatrix relate to gross investment by industries in the products of different industries. They are of three different kinds: (i) gross investment proper in fixed assets, which appears in rows 6, 7, 8, 11 and 13; (ii) legal fees, stamp duties etc., which appear in row 17 and are spread over the columns in proportion to the entries in row 13; and (iii) investment in stocks, which appears diagonally. Thus (6.61) shows not only the engineering, etc. products obtained for gross investment by the engineering, etc., industry but also the increase in stocks and work in progress of that industry's products in whatever industry they may be held. All these estimates are based on the *Blue Book*.

T_{17}. The entries in this submatrix relate to the net foreign demand for the products of British industries. These exports less the corresponding competitive imports all appear in column 85. Competitive imports are items like machinery, cars, clothing and furniture, for which a home industry exists but which for one reason or another are imported. Non-competitive, or complementary, imports are things like raw cotton, most metal ores and crude petroleum, which either cannot be produced in Britain or are not produced here in large quantities. All such imports appear in row 85. Although the following refinement is not used here, the minor British output of these products could conveniently be shown as a negative input from complementary imports into the British producing industry, and the rest of the world production account (row and column 85) could then be put in balance by showing the minor British production as a positive input from complementary imports into the British using industry. All these estimates are obtained from a detailed analysis of trade statistics.

T_{24}. The entries in this submatrix relate to consumers' expenditure on goods and services and all appear in column 50. They include expenditure by foreign visitors as well as by Britons, and the excess of British tourist expenditure abroad over expenditure by foreign visitors in Britain appears in (30.50).

T_{34}. The entries in this submatrix relate to government current expenditure on goods and services classified by purpose. The entries for the central government appear in column 53 and those for local authorities in column 55.

T_{41}. The entries in this submatrix relate to the income of the factors of production engaged in British industries and to those indirect taxes (net) which are treated as a cost of production rather than as a direct charge against a particular class of consumers' expenditure. Income from self-employment, other profits, interest and industrial rental income appear in row 47. Rental income from dwellings appears in row 48. Income from employment appears in row 50. Indirect taxes (net) paid to the central government appear in row 53 and rates paid to local authorities appear in row 55.

T_{42}. The entries in this submatrix relate to consumers' expenditure on direct labour (50.30) and to those indirect taxes which are charged directly against expenditure and are not routed through one of the eighteen industries.

T_{43}. The entries in this submatrix relate to direct labour employed by public authorities in connection with their various purposes. All the entries appear in row 50.

T_{44}. The entries in this submatrix relate to the distribution of non-labour income over institutions (the *Blue Book* sectors) and to current transfers between these sectors. It is perhaps clearest to deal with the non-zero entries in order.

(47.53) is central government debt interest which is paid into account 47. This account distributes non-labour income, other than rental income from dwellings, over sectors.

(47.55) is local authority debt interest other than that paid to the central government.

(49.47) is income, before British direct taxes but after dividend payments and foreign direct taxes, accruing to companies.

(50.47) is income from self-employment, profits, interest and industrial rents accruing to persons.

(50.48) is rental income from dwellings accruing to persons.

(50.53) is income from transfers from the central government accruing to persons.

(50.55) is income from transfers from local authorities accruing to persons.

(51.47) is income, other than rental income from dwellings, accruing to public corporations.

(51.48) is rental income from dwelling accruing to public corporations.

(52.47) is trading income accruing to central government enterprises.

(52.48) is rental income from dwellings accruing to the central government.

(53.47) is income from property, other than interest from local authorities, accruing to the non-trading departments of the central government.

(53.49) is direct taxes on income received by the central government from companies.

(53.50) is direct taxes on income, including national insurance and health contributions, received by the central government from persons.

(53.51) is direct taxes on income received by the central government from public corporations.

(53.55) is interest received by the central government from local authorities.

(54.47) is trading income accruing to local authority enterprises.

(54.48) is rental income from dwellings accruing to local authorities.

(55.47) is income from property accruing to the non-trading departments of local authorities.

(55.53) is current grants received by local authorities from the central government.

T_{51}. In this submatrix non-zero entries occur only in the leading diagonal. These relate to provisions for depreciation in the different industries valued on a replacement cost basis. The estimates are made by allocating to industries the *Blue Book* figures for depreciation in respect of different categories of asset. This allocation is based on gross investment at 1954 prices accumulated over the years 1948 through 1959.

T_{56}. The entries in this submatrix relate to the finance for net investment in fixed assets and stocks in the different industries. As was the case with the distributions of incomes from property, it is impossible to do more than record the marginal totals of a complete to-whom-from-whom classification. The stock figures have already appeared in the leading diagonal of T_{15} and the estimates for net investment in fixed assets are the excess of gross investment over depreciation.

T_{64}. With one exception, the items in this submatrix represent the saving of the different institutional sectors. The exception is (77.47) which represents the residual error. It differs from the figure given in the *Blue Book* for 1960 because revised estimates have been used for the balance of payments. This means that the estimates of property income shown in (47.1) to (47.17) differ correspondingly in total from the estimate in the *Blue Book*.

T_{66}. The entries in rows 74 through 77 of this submatrix represent the expenditure by the different financial sectors on net additions to (i) fixed assets other than dwellings, (ii) dwellings, (iii) stocks and work in progress and (iv) financial claims. The remaining entries represent capital transfers. Thus (82.79) represents death duties and similar taxes on capital, and (78.82), (79.82), (80.82) and (84.82) represent capital transfers from the central government to various financial sectors.

T_{67}. This submatrix contains a single entry, (77.87), which represents net lending by Britain to the rest of the world.

T_{71}. This submatrix contains complementary imports of intermediate products. These imports are here given a wide definition and include not only imports of intermediate products for which a corresponding British industry does not exist, such as copper ore or raw cotton, but imports of all intermediate products for which home production provides less than half the total supply.

T_{72}. This submatrix contains complementary imports of final goods destined for private consumption, such as tropical fruits and foreign wine. (85.30) represents the excess of expenditure abroad by British tourists over similar expenditure by foreigners in Britain.

T_{73}. This submatrix contains complementary imports of final goods destined for government consumption: in fact, overseas expenditure by the central government in connection with foreign services and defence.

T_{74}. This submatrix contains three items: (86.47) represents income from property paid abroad (net); (86.50) represents remittances abroad (net) by persons; and (86.53) represents similar current transfers abroad (net) by the central government.

T_{76}. This submatrix contains a single entry, (87.82), which represents capital transfers received by the rest of the world from the central government.

T_{77}. The items in this submatrix arise because the rest of the world is here represented by three accounts rather than by one. (86.85) represents the rest of the world's balance of trade with Britain and (87.86) represents the rest of the world's balance of current payments with Britain.

3. A PROVISIONAL TABLE FOR 1959

The following table contains a provisional social accounting matrix for Britain in 1959 drawn up on the lines which have just been described. As is usual, money incomings are shown in the rows and money outgoings are shown in the columns so that, as can be seen, each of the 87 accounts balances. This table is the work of the group engaged on the Cambridge Growth Project, namely Mr J. A. C. Brown, Mr J. M. Bates, Mr F. G. Pyatt, Mr M. O. L. Bacharach, Dr Z. Pawlowski and Mlle C. Vannereau, in addition to myself. On behalf of the group I should like to thank the Board of Trade for their kindness in making available to us the input–output data for 1954 now published in [174] and for their help in the interpretation of this material.

XVII

BRITISH ECONOMIC BALANCES IN 1970: A TRIAL RUN ON ROCKET

1. INTRODUCTION

For rather more than three years a group of us at the Department of Applied Economics has been working on a computable model of British economic growth. The aim of this project is to study quantitatively in as great detail as possible the present structure and future prospects of the British economy, the possibility of stimulating its rate of growth, and the problems to which this would give rise. The progress of this work is reported in a series of publications entitled *A Programme for Growth* [30–35].

Before sitting down to write this paper, I checked up on the writings which have so far emerged from this work. I found that in the two and a half years ending in December 1963 we had prepared twenty-nine booklets, articles and conference papers, most of which have already been published. In view of this large amount of explanatory material, I feel justified here in taking as read the general philosophy and objectives of this work as well as many of the technical details of carrying it out. I shall concentrate on the preliminary results for 1970 which have emerged from our trial run, describing only the economic relationships which led to these results and the steps we are taking to improve these relationships.

At the outset I should like to emphasize a number of general points which help to explain the figures given in this paper and show what they are and what they are not.

First, they are preliminary figures, the outcome of a trial run of our computing programme *Rocket*. In reaching them we have made use of a large body of data on consumption patterns, industrial inputs, capital requirements, saving propensities and so on. This material, much of which relates to the future, is capable of refinement and needs to be checked as far as possible by people with a practical knowledge of different parts of the economic system. We are now engaged in this process of improvement.

Second, our figures are not forecasts; they are intended to show possibilities and problems. In this paper I shall examine the consequences of only one set of assumptions, although our aim, as we make the model more accurate, is to examine several alternatives. The assumptions chosen here can claim a particular interest because they lead to the same fundamental growth rate, 4 per cent, as has been adopted by the National Economic Development Council.

Third, our figures do not relate to the immediate future; we have concentrated our first endeavours on the consequences of achieving a certain level of consumption in 1970 and a certain rate of growth in consumption thereafter. The next step is to see what would have to happen during the remainder of the 1960's if the initial conditions for realizing the 1970 picture were to be met. We are working on this problem now and its solution may well cause us to revise our ideas on the state of affairs that can be considered desirable in 1970.

Fourth, as economic models go, ours is comparatively large and detailed. We have built a disaggregated model because we attach importance to the interest and co-operation of people in practical life whose expertise relates to a particular facet of the economy; a man may know a lot about inputs into steel or the market for domestic refrigerators without being able to say much about inputs into metals in general or the market for household appliances as a whole. From this point of view it may be thought that we have not gone nearly far enough: we have only thirty-one industries in our model. There are two answers to this. In the first place, a start must be made: everything cannot be done at once. In the second place, there are objections as well as difficulties in getting more detail into a model simply by making it larger by disaggregation. If we consider, say, the energy or transport complexes among industries or the educational or health services among government activities, it is evident that we should want the detailed treatment to run along the lines of activity analysis rather than the simpler input–output approach we have used. Accordingly, in trying to extend the model in detail, we believe the main task to lie in the development of sub-models which are linked to the main model but represent their facet of the economy in a more sophisticated way. We are at present planning an experiment in the form of a sub-model for the energy complex.

Fifth, our model as it stands at the moment relates mainly to the physical rather than the financial aspects of production, consumption, accumulation and foreign trade. We are now beginning to work on the financial aspects too. Another side of the picture which we have begun to study is the changing spectrum of skills required in different industries. Gradually, as the work proceeds we are led into fields, such as the pattern of education and training, the impact of research and development, and attitudes to innovation, more and more remote from the area of pure economics.

I have perhaps said enough to show the organic character of our work. Our model is a working model but it is still at the exploratory stage, as will be seen from the description I shall give of it. We hope and believe that before long it will be very much better.

2. THE FRAMEWORK OF THE MODEL

It may be comforting to a traveller about to embark on an arduous journey to a strange land to be shown a glimpse of what he may expect to find at his journey's end. If the picture is not too amazing, he may be encouraged to feel that he will be able to live in the strange land; perhaps it will not even seem very strange. With this in mind I propose to begin by presenting in table 1, on p. 252, a highly consolidated social accounting matrix for 1960, taken from [31], and its counterpart for 1970.

The social accounting matrix of the model consists of 253 accounts; in table 1 these are reduced to 13 consolidated accounts. All entries are at 1960 values and, as in input–output tables, incomings (or revenues) are shown in the rows and outgoings (or costs) are shown in the columns. The accounts are shown in four groups, 1 through 4, 5 and 6, 7 through 14, and 15. If the accounts within each group are consolidated we obtain the four national accounts relating to production, consumption, accumulation and the rest of the world.

In [31] details of the definitions, sources and methods used in constructing the estimates for 1960 are set out in full. Here I shall only outline the meaning of the accounts and entries, using the 1970 figures for the purpose.

The first account relates to commodities or products. The entries in the first column show where these products come from; those in the first row show where they go to. The first entry in column 1, £60363 million, shows the amount that comes from British industry. The remainder comes from abroad in the form of competitive imports, and is shown in two parts: £3456 million are the payments made abroad for these imports, and £207 million are the customs duties payable on them at 1960 rates. If we add these three entries together we obtain the total value of commodities available, £64026 million.

The destinations of these commodities are shown in the first row of the table. Thus £28737 million go to industries in the form of intermediate products needed for current production. A further £17220 million and £2128 million go respectively to private and government consumers for the purpose of consumption. The usage of commodities for capital purposes is shown next. Thus £319 million go to increase stocks and work in progress; £1225 million and £3196 million go respectively for purposes of replacing and extending industrial fixed assets; £3324 million go to replacing and extending consumers' stocks of durable goods and dwellings; and £726 million go to replacing and extending social capital in the form of schools, hospitals, roads and the like. Finally, the last entry in the row, £7151 million, goes abroad in the form of British exports. In total, these uses exhaust the supply of commodities, £64026 million.

The second account relates to industries, in which the commodities are produced. If we look at row 2 we see that industries sell their total output,

TABLE 1 SOCIAL ACCOUNTING MATRICES FOR BRITAIN: 1960 AND 1970

(1960 £ million)

Column groups — Production accounts: serials 1–4; Income and outlay accounts: serials 5–6; Capital transactions accounts: serials 7–14; All accounts: serial 15; Total incomings: serials 1–15. In each cell the upper figure is 1960 and the lower figure is 1970.

Type of account	Description of class	Serial number of class	No. of accounts in class	1 (31)	2 (31)	3 (40)	4 (12)	5 (6)	6 (6)	7 (25)	8 (31)	9 (31)	10, 11 (12)	12, 13 (16)	14 (9)	15 (3)	1–15 (253)
Production accounts	Commodities	1	31		19329 / 28737	11917 / 17220	1528 / 2128			563 / 319	1065 / 1225	1845 / 3196	1916 / 3324	338 / 726		4751 / 7151	43252 / 64026
	Industries	2	31	40652 / 60363													40652 / 60363
	Consumers' goods and services	3	40						16214 / 23562							215 / 320	16429 / 23882
	Government purposes	4	12						4189 / 5963								4189 / 5963
Income and outlay accounts	Indirect taxes and subsidies	5	6	135 / 207	488 / 802	1946 / 2917	44 / 63			1 / 0	32 / 52	45 / 172	225 / 429				2916 / 4642
	Institutional sectors	6	6		17056 / 25314	728 / 777	2338 / 3322	2916 / 4642								179 / 250	23217 / 34305
Capital transactions accounts	Commodities, additions to stocks	7	25												591 / 333		591 / 333
	Industries, replacements	8	31		1097 / 1277												1097 / 1277
	Industries, extensions	9	31		489 / 1256										1401 / 2112		1890 / 3368
	Consumers' goods, replacements and extensions	10, 11	12			1265 / 2188									876 / 1565		2141 / 3753
	Government purposes, replacements and extensions	12, 13	16				133 / 256								205 / 470		338 / 726
	Institutional sectors	14	9						2728 / 4680								2729 / 4680
All accounts	Rest of world	15	3	2465 / 3456	2193 / 2977	573 / 780	146 / 194		86 / 100	27 / 14					-344 / 200	1 / 0	5146 / 7721
	Total outgoings	1–15	253	43252 / 64026	40652 / 60363	16429 / 23882	4189 / 5963	2916 / 4642	23217 / 34305	591 / 333	1097 / 1277	1890 / 3368	2141 / 3753	338 / 726	2729 / 4680	5146 / 7721	

Main sources: for 1960 (upper figures), [31]; for 1970 (lower figures), computer run 20029.

£60363 million, to the commodity account which distributes this output, together with competitive imports, to the various users.

If we look at column 2 we see in broad outline the cost structure of British industry. We have already noticed the £28737 million of intermediate product which consists of domestic products and competitive imports; to this we must add from the last entry in column 2 a further £2977 million of intermediate product which comes from abroad in the form of complementary imports. The remaining items of cost are as follows: indirect taxes net of subsidies, £802 million; income payments of all kinds, £25 314 million; and industrial depreciation divided between £1277 million needed to finance replacements of fixed assets and a further £1256 million available to finance extensions. All these costs add up, of course, to £60363 million, the total sales proceeds of British industry.

Why, it may be asked, do we distinguish between industries and their products? The reason is that there is not a one-to-one correspondence between the two concepts; and since most of our industrial data come in the form of an industry's make of various products and of its absorption of various products, it is convenient to set up the accounting system to contain this information. For example, in our calculations for 1970, the total British output of engineering products is £5787 million of which £5462 million is made in the engineering industry and £325 million in other industries. Correspondingly, the total output of the engineering industry is £5821 million of which £5462 million consists of engineering products, as above, and the remaining £359 million consists of products characteristic of other industries.

The third account relates to consumers' goods and services. There are two incoming entries: £23562 million in respect of personal consumption and £320 million in respect of the consumption expenditures of foreign visitors. From the column we see that most of this money goes in the purchase of commodities: £17220 million on domestic products and competitive imports; and a further £780 million on complementary imports into private consumption, such as tropical fruit, tea, cigars and holidays abroad. The remaining outgoings relate to: indirect taxes (net), £2917 million; income payments, for example to landlords and domestic servants, £777 million; and depreciation on consumers' durables and dwellings, £2188 million. In our model not only do we treat dwellings as assets, the usual practice, but we do the same with consumers' durables such as furniture, domestic appliances and vehicles.

The fourth account relates to government consumption, which is classified by purpose. The single incoming entry, £5963 million, appears at the intersection of row 4 and column 6. The cost components of this total, shown in column 4, are analogous to those already described for private consumers.

This brings me to the end of my description of the four classes of production account that appear in our model. The need for classes 3 and 4 arises because, just as there is no one-to-one correspondence between industries and commodities, so there is no one-to-one correspondence between commodities and either consumers' goods and services or government purposes. But in order to work out the industrial implications of consumers' or government demands we must be able to convert these demands, expressed in their own classifications, into demands for commodities and thence into activity levels in different industries.

We come now to the income and outlay accounts, of which there are two. The first, account 5, relates to indirect taxes and subsidies. Its purpose is a simple one: to collect together these items wherever they appear in the system and pay them into the institutional income and outlay account. Most of the incoming items in row 5 we have met before; the exceptions are the taxes debited to the capital transactions accounts which represent purchase taxes on vehicles and on household durables.

The sixth account is the income and outlay account of all institutional sectors, private and public, in the economy. Apart from the £4642 million of indirect taxes (net of subsidies) which flow in from account 5, all other incomings represent receipts of income to the factors of production. The total of these receipts is simply the national income, £29 663 million. Of this total: £25 314 million represents income payments by industry; £777 million represents income payments made directly by persons; £3322 million represents income payments made directly by central and local government in the form of wages and salaries; and £250 million comes from abroad in the form of net income on foreign investments. When net indirect taxes are added, the grand total is £34 305 million.

The outlay of this total is shown in column 6. Thus £23 562 million represents private consumption; £5963 million represents government consumption; £4680 million represents saving; and £100 million represents net current transfers abroad.

Accounts 7 through 13 relate to expenditures on stocks and fixed assets and need little explanation. The incoming items relate either to provisions for depreciation, which are debited to the various production accounts, or to the finance of net investment, which is debited to the capital transactions account of institutions. The outgoing items relate mainly to the purchase of commodities from account 1. In addition there are purchase taxes paid to account 5 in respect of vehicles and household durables and an entry at the intersection of row 15 and column 7 which relates to the increase in stocks of complementary imports into industry.

The fourteenth account is the capital transactions account of institutions and brings together the flows involved in financing the increase in the national wealth. The single item flowing into this account comes from the income and outlay account of institutions and represents the saving of all sectors of the economy, £4680 million. The outgoing items relate to domestic net investment in stocks and work in progress, £333 million, in industrial fixed assets, £2112 million, in dwellings and consumers' durables, £1565 million, and in social capital, £470 million, and to net lending abroad, or foreign investment, £200 million. The sum of these items represents the net addition to the country's wealth and is equal to its saving.

The fifteenth and last account relates to the rest of the world. The incoming items are the proceeds of sales to Britain and gifts and borrowing from Britain. The outgoing items are purchases from and income payments to Britain.

This is the consolidated framework of the model; if we go back to the original 253 accounts we have the framework itself. The source for the 1960 figures is [31]. The source for the 1970 figures is indicated in table 1 as computer run 20029. Let us now see what enabled the computer to produce these figures.

3. EXOGENOUS FINAL DEMAND

The starting point of the calculations is the estimation of exogenous final demands in 1970 for each of the thirty-one commodities or products shown separately in the model and linear growth rates for these demands assumed to hold from 1970 onwards.

The components of exogenous final demand are six in number: private consumption, government consumption, industrial replacements of fixed assets, investment in dwellings and consumers' durables, investment in social capital, and exports. These components are shown separately and in total in table 2 together with the aggregate changes.

The figures in this table are estimated demands for commodities. In the case of exports, estimates can be made on this basis in the first instance; in all other cases it is necessary to make the initial estimates in terms of another classification and then convert them into demands for commodities. Let us now see how this is done in each case.

(*a*) *Private consumption.* We start by assuming that between 1960 and 1970 private consumers' expenditure as a whole, including net investment in consumers' durables, reckoned at 1960 prices will grow by 47 per cent in all. This is equivalent to a growth rate of 3·9 per cent, or 3·2 per cent per head of the population. A small part of this growth rate is attributable to the growth of net investment in consumers' durables which we shall come to under (*d*) below. If this is left out, the remainder, private consumption proper, is assumed to grow at 3 per cent per head.

The next step is to divide the assumed total of consumers' expenditure among the forty goods and services which appear in the model. This is done in two stages. First, demand equations based on the experience of the years 1900 to 1960 are used to divide total expenditure between eight main groups as explained in detail in [160]. Second, estimates are made of the components

TABLE 2 EXOGENOUS FINAL DEMAND AND ITS RATE OF CHANGE: BRITAIN 1970 (1960 £ million)

Characteristic products of	Private consumption	Government consumption	Industrial replacements	Dwellings and consumers' durables	Investment in social capital	Exports	Total	Annual change in exogenous final demand
1. Agriculture, forestry, fishing	1898	29	0	0	0	107	2034	45
2. Coal mining	193	16	4	0	0	69	282	2
3. Mining and quarrying n.e.s.	22	0	0	0	0	33	55	4
4. Food processing	2246	34	0	0	0	146	2426	33
5. Drink and tobacco	719	0	0	0	0	167	886	25
6. Coke ovens, etc.	35	31	0	0	0	9	75	3
7. Mineral oil refining	358	97	0	0	0	50	505	37
8. Chemicals n.e.s.	395	138	0	0	0	818	1351	79
9. Iron and steel (m. r. and c.)	0	0	0	−29	0	240	211	10
10. Iron and steel (t. and t.)	0	0	0	0	0	116	116	1
11. Non-ferrous metals	−39	0	0	0	0	182	143	6
12. Engineering and electrical goods	170	268	434	444	55	1316	2687	137
13. Shipbuilding and marine engineering	0	124	81	0	0	81	286	6
14. Motors and cycles	0	41	259	657	34	1173	2164	99
15. Aircraft	0	278	10	0	0	259	547	22
16. Railway locos. and rolling stock	0	0	73	0	0	3	76	0
17. Metal goods n.e.s.	130	42	11	27	3	207	420	13
18. Textiles	230	14	0	77	0	284	605	9
19. Leather, clothing, footwear	1714	10	0	0	0	85	1809	134
20. Building materials	0	16	0	0	0	25	41	0
21. Pottery and glass	76	11	0	0	0	55	142	4
22. Timber, furniture, etc.	0	22	7	190	13	19	251	13
23. Paper, printing, publishing	310	24	0	0	0	166	500	12
24. Other manufacturing	199	45	0	42	0	134	420	19
25. Construction	379	276	218	1113	619	10	2615	117
26. Gas	122	24	1	0	0	4	151	−1
27. Electricity	693	60	21	0	0	6	780	42
28. Water	72	13	5	0	0	12	102	2
29. Transport and communications	1092	52	54	24	0	836	2058	35
30. Distributive trades	3893	64	2	742	2	16	4719	252
31. Services n.e.s.	2313	399	45	37	0	523	3317	168
Total	17220	2128	1225	3324	726	7151	31774	1328

of these groups by reference to their changing relative importance within the group; for example, within the food group, the proportion spent on bread and cereals tends to fall with time, whereas the proportion spent on meat, fruit and vegetables tends to rise. We have also tried to allow for substitutions, such as the substitution of electricity and oil for coal as a fuel.

We now have consumers' expenditure in 1970 divided into forty categories. We then remove the estimated expenditure on the five categories of consumers' durables and replace this expenditure by a consumption component represented by depreciation. The resulting total, as we can see from table 1, is £23 882 million.

This total and its forty components must now be converted into demands for commodities and into the other demands shown in the third column of table 1. This calculation is based on the experience of 1960 contained in [31]. In this way we divide consumers' expenditure on clothing, for example, into demands for textiles, clothing products, rubber goods (included in other manufacturing), transport and distribution services and purchase taxes at 1960 rates. The figures in the first column of table 2 are obtained by adding up for each industrial product the components for each of the forty consumer goods and servides.

We thus see that the figures in column 1 of table 2 depend essentially on three factors: the initial assumption about the level of total consumers' expenditure to be reached in 1970; the use of demand equations to divide this total into the forty consumers' categories; and finally, the conversion of these categories into demands for commodities. We thus see that there are three steps involved in the calculation: the first is a pure assumption and can be varied at will; the second results from the analysis of past observations and is divided into two stages; and the third is based on the classification converter for 1960 embodied in [31].

The second and third steps in the calculation are based on econometric analysis which, in principle, can always be checked against the impressions of people engaged in practical activities: how, for example, do our estimates of the future demand for clothing or cars or cinemas check with the views of the producers and distributors of these goods and services or with the views of other economists engaged in making similar projections ? So far we have not reached the stage of looking for this kind of check; we have been occupied with the econometric analysis of published data. Let us now see how far we have got and what improvements could be made.

As far as demand analysis is concerned, our methods and results for the eight main groups are set out in detail in [160] which also contains a description of the projections for 1970. The computing programme used, which allows for linear trends in the parameters, and a further programme which allows for quadratic trends [XIV], involve an iterative two-stage least squares routine that requires a very large number of calculations. The analysis of group components can be fitted into this system but, as explained above, this has not yet been done.

At the moment we are engaged in improving our demand projections in several ways: by elaborating the equations; by altering the computing method to enable all the groups to be analysed simultaneously; and by combining budget material with time series in estimating the demand parameters. I shall now explain the steps we are taking.

The demand model at present in use can be written in the form

$$\hat{p}e = b\mu + (I - bi')\hat{a}p$$

$$= \hat{p}a + b(\mu - p'a) \tag{1}$$

$$a_\theta = a^* + \theta a^{**} \tag{2}$$

and

$$b_\theta = b^* + \theta b^{**} \tag{3}$$

The meaning of these symbols is as follows. The symbol p denotes a vector of prices and e denotes a vector of quantities bought. A circumflex indicates that a diagonal matrix is formed out of a vector and so $\hat{p}e$ denotes a vector of expenditures. The symbol μ denotes total expenditure so that $\mu \equiv p'e$ where the prime indicates transposition: in this case a column vector is transposed into a row vector. The symbols a and b denote vectors of constants restricted by the fact that the sum of the elements of b is one: $i'b = 1$ where i is the unit vector. The symbol I denotes the unit matrix and θ denotes a particular year.

From the second row of (1) we see that according to this model consumers buy an amount of each commodity represented by an element of a. The total cost of this is $p'a$ and so $\mu - p'a$ denotes the amount of money left over when they have made these purchases. This amount of money they spread over the different commodities in proportion to the elements of b.

From (2) and (3) we see that the elements of a and b are linear functions of time, a^*, a^{**}, b^* and b^{**} being vectors of parameters which satisfy the constraints $i'b^* = 1$ and $i'b^{**} = 0$. Thus if we work throughout with per head figures we can interpret the elements of a, which change through time, as the components of the average consumer's basic standard of living and we can interpret the elements of b, which also change through time, as the average consumer's allocation coefficients for uncommitted expenditure.

The method of calculation we are now proposing to use enables us, in principle, to analyse, in a single calculation, any number of consumer categories, provided only that these are so chosen that complementary and inferior groups do not appear. The method is as follows.

If we premultiply (1) by $\hat{p}^{-1}$, the diagonal matrix of price reciprocals, we obtain

$$e = a + \hat{b}[\hat{p}^{-1}i(\mu - p'a)]$$

$$= a + \hat{b}y \tag{4}$$

say, where the elements of y are total uncommitted expenditure expressed in the prices of one of the consumers' goods and services. We can form an initial estimate of y either by putting $a = \{0, 0, \ldots, 0\}$ or, better still, by using the values of a gained from an earlier analysis. We then fit (4) subject to the constraint $i'b = 1$, obtain estimates of a and b, revise y and continue until convergence is reached.

This system of equations has many convenient properties described in [30, 153, 160] which I shall not repeat here. The computational method suggested above can readily be generalized. Thus if we wish to incorporate (2) and (3) in the model, we simply obtain

$$e_\theta = a^* + \theta a^{**} + \hat{b}^* y_\theta + \hat{b}^{**} \theta y_\theta \tag{5}$$

to be fitted subject to $i'b^* = 1$ and $i'b^{**} = 0$.

Experience shows that linear or quadratic trends in the parameters, though useful as a first approximation, have disadvantages for purposes of projection. A better course, therefore, might be to make a and b functions of past values of consumption. For example, let the elements of e_θ^* denote, say, three-year moving averages of the consumption of different goods and services in the years immediately preceding θ. Then we could replace (2) and (3) by

$$a_\theta = a^* + \hat{e}_\theta^* a^{**} \tag{6}$$

and

$$b_\theta = b^* + i' e_\theta^* b^{**} \tag{7}$$

so that in place of (5) we should have

$$e_\theta = a^* + \hat{a}^{**} e_\theta^* + \hat{b}^* y_\theta + \hat{b}^{**} i' e_\theta^* y_\theta \tag{8}$$

to be fitted, again, subject to $i'b^* = 1$ and $i'b^{**} = 0$. The advantage of this form is that the parameters neither change at a constant rate nor do they pass through maxima or minima unless the consumption series do the same.

Another problem that must be faced explicitly, largely because of the increasing importance of durable goods, is the introduction of adaptive behaviour into the model. In the equations considered so far, consumers are constantly in a state of equilibrium, given the money they have to spend and the prices with which they are faced. In practice they may take some time to adapt themselves to changing circumstances. In this case, as shown in [30, 160], we must replace (1) by the more elaborate form

$$\hat{p}e = \hat{c}b\mu + \hat{c}(I - bi')\hat{a}p + (I - \hat{c})\hat{p}x$$
$$= \hat{c}\hat{p}a + \hat{c}b(\mu - p'a) + (I - \hat{c})\hat{p}x \tag{9}$$

where c denotes a vector of the ratios of adjustment rates to depreciation rates and x denotes a vector whose elements represent consumption in the preceding year in the case of perishable goods (depreciation rate equal to one) and a calculable function of the stock at the beginning of the year in

18

the case of durable goods (depreciation rate less than one). Corresponding to (4) we now have

$$e = \hat{c}a + \hat{c}\hat{b}y + (I - \hat{c})x \tag{10}$$

to be fitted subject to $i'b = 1$.

We have found [153] that uncommitted expenditure, $\mu - p'a$, is usually very small, of the order of 10 to 15 per cent of income. The substitution effects of prices are therefore very small. A price sensitive variant of (1) which would preserve its convenient theoretical properties would take the form

$$\hat{p}e = b\mu + (I - bi')(\hat{a} + \hat{p}A)p$$
$$= \hat{p}(a + Ap) + b[\mu - p'(a + Ap)] \tag{11}$$

where A is a symmetric matrix. Corresponding to (4) we now have

$$e = a + Ap + \hat{b}\{\hat{p}^{-1}i[\mu - p'(a + Ap)]\}$$
$$= a + Ap + \hat{b}y \tag{12}$$

say, to be fitted subject to $A = A'$ and $i'b = 1$.

The equations of the models considered so far have to be fitted simultaneously since there are one or more constraints on their parameters. In principle there is no great difficulty in this but in practice the method may not work well, partly because of economic interdependence and partly because of difficulties that arise with large near-singular matrices. We can, however, begin by grouping the series and proceed as follows.

Let us apply a grouping matrix, G say, to (1). Then since $i'G = i'$, we have

$$G\hat{p}e = Gb\mu + G(I - bi')\hat{a}p$$
$$= Gb\mu + (I - Gbi')G\hat{a}p$$
$$= Gb\mu + (I - Gbi')(G\hat{a}G')\bar{p} \tag{13}$$

which is of exactly the same form as the complete system provided that

$$\bar{p} = (G\hat{a}G')^{-1}G\hat{a}p \tag{14}$$

that is, if the price series for the groups are weighted averages of the component prices with the a_j as weights.

Thus we can start with the sub-groups and obtain a full range of a_j. We can calculate the elements of b for the main groups, $_jb$ say, as

$$_jb = p_j'(e_j - a_j)/(\mu - p'a) \tag{15}$$

and provided that we define the group price index-numbers as in (14) we shall find that everything is consistent. This idea, which was suggested to me by D. A. Rowe, seems likely to prove extremely useful in practice.

In addition to this work on time series, we have recently turned our attention to budget data and have compiled information for our forty groups from the official budget surveys of 1937–38, 1953–54, 1959, 1960,

1961 and 1962. Experience shows [XIV] that in the case of food and clothing, the two groups we have examined, there is a considerable degree of consistency between expenditure elasticities derived from the model and from budgets. If this proves to be the case generally, then we can greatly improve our estimates of the demand parameters by combining budgets and time series.

In this long digression on demand analysis, we have seen the importance of allowing for changes in preferences as summarized in the vectors a and b. We may also expect that gradual changes take place in the relationship between particular goods and services and the demands for commodities to which they give rise. Our intention here is to construct classification converters for years other than 1960 and in this way form an idea of how the coefficients are changing.

In concluding this account of the estimation of private consumption I must say a word about future prices. It is only relative prices that matter, but if these are expected to change in relation to 1960 we must adjust the amount of money needed to keep the average consumer at the level of utility intended. We have so far obtained 1970 prices simply by extrapolating price trends but eventually these estimates will be replaced by shadow prices calculated from the complete model.

(b) *Government consumption.* We assume that between 1960 and 1970 government consumption will grow by 42 per cent in all or at a rate of 3·5 per cent, and that its components will continue to move along the trends indicated by the National Economic Development Council. In terms of the main components this means that consumption expenditure connected with education will grow considerably more than the average, consumption expenditure connected with health will grow a little more than the average and that expenditure on defence and other current purposes will grow rather less than the average.

Once we have these estimates for the twelve purposes into which government consumption is divided in the model, the next thing to do is to pass them through a classification converter and turn them into demands for industrial products. The results are shown in the second column of table 2. The large figures associated with engineering, shipbuilding, vehicles and aircraft are mainly due to the fact that, following the usual convention in social accounting, we have treated all defence expenditure as current.

It would now be possible to improve these estimates of government consumption since the Treasury has started to publish projections of central government expenditure. The first publication relates to 1967–68 and provides a considerable amount of detail, but this information was not available when the present calculations were made.

(c) *Industrial replacements.* At present our estimates of industrial replacements are based on the assumption that all fixed assets are automatically replaced as they reach the end of their assigned life-spans, different life-spans naturally being assigned to different assets. Future expenditure on replacements is thus made to depend on these life-spans and on the past history of investment expenditures. The figures in the third column of table 2 represent not the replacement programmes of the different industries but the demands for different products to which these programmes, in the aggregate, give rise. Naturally, these demands are concentrated fairly heavily on engineering, vehicles and construction.

We are well aware that this is not an ideal way of estimating industrial replacements and that scrapping and replacement decisions should be made to depend on economic circumstances and not to flow automatically from what has happened in the past. We are now working on a form of production function which will enable us to make this improvement. I shall not discuss the matter here, however, since these functions are the subject of a separate paper by my colleague Graham Pyatt [129].

(*d*) *Consumers' durables and dwellings.* We have already seen that expenditure on consumers' durables has been estimated as part of the calculations on consumers' expenditure. Again it is necessary to convert the expenditures into demands for industrial products and to add in a figure for gross investment in dwellings, which we have assumed to be 48 per cent above the 1960 level. We see from the fourth column of table 2 that in addition to large demands on engineering, vehicles, furniture and construction there is also a large demand for distributive services associated with consumers' durables. We also see that there is a negative demand for iron and steel, just as there was a negative demand for non-ferrous metals in the first column of the table. These represent an attempt to allow for scrap metal which adds to industrial supplies. This scrap comes mainly from cars in the case of iron and steel and mainly from domestic utensils and fittings in the case of non-ferrous metals.

(*e*) *Social capital.* We assume that that part of public expenditure which we term gross investment in social capital, namely educational buildings, hospitals, roads and so on, will more than double between 1960 and 1970. This rise is mainly due, on our assumptions, to an increase of 88 per cent in educational facilities and of 200 per cent in roads and street lighting. As we can see from the fifth column of table 2, this expenditure gives rise mainly to a demand for construction.

(*f*) *Exports.* We assume that between 1960 and 1970 the demand for British exports will grow by 51 per cent in all, or at a rate of 4·1 per cent. This assumption is based on estimates carried out for individual product-groups. In effect, we are relying on chemicals, oil, vehicles, aircraft, metals and metal goods for the largest proportionate increases; it is from these together with engineering products that we expect to get the bulk of our increased earnings from foreign trade. We expect services to rise by less than the average and we expect textiles and railway equipment actually to decline.

(*g*) *Total exogenous final demand and its rate of change.* If we form the row totals of the elements in the preceding six columns we obtain the components of exogenous final demand as a whole. These are shown in the seventh column of table 2.

If we calculate the changes assumed to be taking place in the exogenous components of final demand that emerge from all the foregoing calculations, we obtain the figures in the eighth and last column of table 2. These estimates are based on the assumptions that private consumption will grow at a rate of 4 per cent per head after 1970 and that all the other items will continue along their existing trends. As we shall soon see, these figures are needed to calculate industrial extensions which have to provide for the increase in capacity from one year to the next. They do not, of course, represent the whole of the increase in the demand for products any more than exogenous final demand is the only kind of demand. We must now turn to endogenous demand, which is composed of: demand for intermediate products, to make current production possible; demand for additional stocks and work in progress, to keep the pipe-lines at a normal level; and demand for industrial extensions, to provide for future increases in industrial capacity.

Once we have taken these endogenous demands into account, we can calculate the first set of economic balances: the accounts for the supply and demand of industrial products.

4. ENDOGENOUS DEMANDS AND COMMODITY BALANCES

The way in which we have calculated endogenous demands in 1970, and so been able to estimate the commodity balances in that year which are implied by our assumptions about exogenous final demands, can best be explained algebraically. In what follows I shall describe the variant of the model we have used in reaching the results set out in table 3. I shall make certain simplifications: complementary imports into consumers' goods and services and government purposes will be ignored and the quantity units will be so chosen that all prices can be put equal to one.

We start with the simple input–output flow equation

$$q = Aq + f \tag{16}$$

where q denotes a vector of domestic commodity outputs, f denotes a vector of final demands for commodities and A denotes a commodity–commodity input–output matrix. The vector f is divided into an exogenous part, f^* say, and an endogenous part, f^{**} say, so that

$$f \equiv f^* + f^{**} \tag{17}$$

TABLE 3   COMMODITY BALANCES: BRITAIN 1970

(1960 £ million)

Characteristic products of	Demands										Supplies		
	Intermediate demand	Private consumption	Government consumption	Additions to stocks	Industrial replacements	Industrial extensions	Dwellings and consumers' durables	Investment in social capital	Exports	Total	Domestic	Foreign	Total
1. Agriculture, forestry, fishing	267	1898	29	5	0	0	0	0	107	2306	1960	346	2306
2. Coal mining	683	193	16	−5	4	39	0	0	69	1000	998	1	1000
3. Mining and quarrying n.e.s.	177	22	0	−1	0	0	0	0	33	233	204	29	233
4. Food processing	807	2246	34	4	0	0	0	0	146	3238	2660	578	3238
5. Drink and tobacco	18	719	0	8	0	0	0	0	167	912	851	61	912
6. Coke ovens, etc.	193	35	31	3	0	0	0	0	9	270	269	2	270
7. Mineral oil refining	505	358	97	2	0	0	0	0	50	1012	905	107	1012
8. Chemicals n.e.s.	2228	395	138	44	0	0	0	0	818	3624	3257	366	3624
9. Iron and steel (m. r. and c.)	2471	0	0	29	0	0	−29	0	240	2711	2685	26	2711
10. Iron and steel (t. and t.)	378	0	0	3	0	0	0	0	116	498	495	3	498
11. Non-ferrous metals	1027	−39	0	13	0	0	0	0	182	1182	1122	61	1182
12. Engineering and electrical goods	2106	170	268	56	434	1381	444	55	1316	6229	5787	442	6229
13. Shipbuilding and marine engineering	103	0	124	0	81	76	0	0	81	465	444	21	465
14. Motors and cycles	665	0	41	38	259	590	657	34	1173	3457	3290	167	3457
15. Aircraft	480	0	278	4	10	9	0	0	259	1040	979	61	1040
16. Railway locos. and rolling stock	84	0	0	0	73	68	0	0	3	228	228	0	228
17. Metal goods n.e.s.	1859	130	42	18	11	34	27	3	207	2331	2250	81	2331
18. Textiles	2560	230	14	19	0	0	77	3	284	3184	2933	251	3184
19. Leather, clothing, footwear	114	1714	10	28	0	0	0	0	85	1951	1807	145	1951
20. Building materials	554	0	16	6	0	0	0	0	25	601	585	16	601
21. Pottery and glass	149	76	11	1	0	0	0	0	55	292	253	39	292
22. Timber, furniture, etc.	627	0	22	12	7	22	190	13	19	913	807	106	913
23. Paper, printing, publishing	1662	310	24	14	0	0	0	0	166	2177	2005	172	2177
24. Other manufacturing	770	199	45	14	0	0	42	0	134	1204	1122	82	1204
25. Construction	1005	379	276	0	218	684	1113	619	10	4304	4304	0	4304
26. Gas	385	122	24	0	1	2	0	0	4	538	521	17	538
27. Electricity	733	693	60	0	21	67	0	0	6	1580	1580	0	1580
28. Water	14	72	13	0	5	4	0	0	12	121	121	0	121
29. Transport and communications	1848	1092	52	0	54	50	24	2	836	3956	3929	28	3956
30. Distributive trades	758	3893	64	0	2	4	742	0	16	5481	5473	8	5481
31. Services n.e.s.	3511	2313	399	0	45	163	37	0	523	6991	6541	450	6991
Total	28738	17220	2128	319	1225	3196	3324	726	7151	64026	60363	3663	64026

Main source: computer run 20029. Components do not always add up to totals because of rounding-off errors.

The vector f^* is composed of two main parts: a domestic component, h say, and a foreign component, x say. Thus

$$f^* \equiv h+x \tag{18}$$

Similarly, the vector f^{**} is composed of two main parts: additions to fixed assets and stocks, v say; and a negative item, m say, which represents competitive imports. Thus

$$f^{**} \equiv v-m \tag{19}$$

Let us stop at this point and consider the changes in these variables from one year to the next. If we multiply (16) by the first-difference operator Δ and substitute from (17) and (19), we obtain

$$\Delta q = A\Delta q+\Delta f^*+\Delta v-\Delta m \tag{20}$$

We must now decide how to calculate Δv and Δm on the right-hand side of (20). The elements of the vector v consist of the commodities needed to provide for an increase from one year to the next either in the fixed assets required for industrial extensions or in additions to stocks and work in progress. Both of these changes depend on the change in domestic outputs, and so we can write

$$v = (K+\hat{k})\Delta q \tag{21}$$

where K denotes a matrix of capital input–output coefficients which connect additional requirements for particular fixed assets with increments of different outputs and where k denotes a vector whose elements relate the additional stocks of a particular product to the additional output of that product. For linear growth in the elements of q, $\Delta^2 q=\{0,0,\ldots 0\}$ and therefore $\Delta v=\{0,0,\ldots,0\}$ too. Evidently, an assumption of linear growth in the elements of q can be used as an approximation to exponential growth.

This brings us to the calculation of Δm. Let us first write down the equations for m, since from these we can obtain what we need by applying the operator Δ to each constituent variable.

This step involves three additional equations. First, we assume that there is a given value for the balance of trade, or excess of exports over imports, β say. That is

$$\beta \equiv i'(x-n-m) \tag{22}$$

where the elements of the vector n denote complementary imports into the different branches of production. Second, we assume that these complementary imports are proportional to the outputs into which they enter. That is

$$n = \hat{a}_3 q \tag{23}$$

where the elements of the vector a_3 denote the factors of proportionality. Third, the various kinds of competitive import, the elements of m, are linear functions of the residual amount of money, $i'm$, available for competitive imports as a whole. That is

$$m = a_1 + a_2 i' m \tag{24}$$

where a_1 and a_2 denote vectors of constants.

From (22), (23) and (24), we see that

$$m = a_1 + a_2(i'x - a_3'q - \beta) \tag{25}$$

If, now, we put $\Delta\beta = 0$, it follows from (25) that

$$\Delta m = a_2(i'\Delta x - a_3'\Delta q) \tag{26}$$

and so, remembering that for linear growth $\Delta v = 0$, it follows from (20) that

$$\Delta q = A\Delta q + \Delta f^* - a_2(i'\Delta x - a_3'\Delta q)$$
$$= (I - A - a_2 a_3')^{-1}[\Delta h + (I - a_2 i')\Delta x] \tag{27}$$

from the definition in (18).

From (21) and (27) we can now express v in terms of h and x, and so we can write

$$q = Aq + f$$
$$= Aq + h + x + v - m \tag{28}$$
$$= (I - A - a_2 a_3')^{-1} \{h + (I - a_2 i')x$$
$$+ (K + \hat{k})(I - A - a_2 a_3')^{-1}[\Delta h + (I - a_2 i')\Delta x] - a_1 + \beta a_2\}$$

This equation enables us to express q in terms of the variables, h, x, Δh, Δx and β, and the parameters, A, K, k, a_1, a_2 and a_3. Once we have determined q in this way, we can estimate v from (21) and m from (25). Intermediate demand is simply Aq. Thus we have all the components to set out the commodity balances as in table 3.

With one exception the column totals of this table have all made their appearance in table 1. The exceptional item is foreign supplies, that is competitive imports, here reckoned inclusive of customs duties which are shown separately in table 1. An exactly similar table for 1960 can be obtained with very little difficulty from table 1 of [31].

I shall conclude this section with a few remarks on the estimation of the parameters used in constructing the table.

I mentioned that the coefficient matrix A relates to the input of commodities into commodities, and so does not arise directly from the system of accounts. At the intersection of row 1 and column 2 of table 1, we have an entry of £28737 million which represents the total of intermediate product absorbed by industries. Let us call the submatrix which would occupy this position in the complete accounting system an absorption matrix, and denote it by X. At the intersection of row 2 and column 1 of table 1, we have an entry of £60363 million which represents the total value of commodities made in all British industries. Let us call the sub-matrix which would occupy this position in the complete accounting system a make matrix, and denote it by M. Then a first approximation to the coefficient matrix A is given by

$$A = X(M')^{-1} \tag{29}$$

In practice, lack of homogeneity gives rise to anomalous coefficients in A which have to be removed. The problems involved are described fully in [32].

For present purposes we considered it necessary to bring existing input–output information up to date and to project it into the future. The way in which this was done is also described in [32]. The matrix used in this paper is in fact an estimate which relates to 1966. We had already made such a projection when this work was undertaken and did not want to make a further projection until we had been able to discuss the results for 1966 with a number of industrial experts. This work is now proceeding and will eventually lead to revised projections.

The matrix K, which relates to the products embodied in the fixed assets needed for additional production of different kinds, is based largely on post-war experience. Much of the basic information is given in [33] but in a number of cases adjustments were made in the light of changes believed to be taking place.

The diagonal matrix $\hat{k}$, which relates to the additional stocks and work in progress needed to ensure the steady flow of production, is again based on post-war experience. This matrix is diagonal because the input of intermediate products into production relates to actual use rather than to purchases. The consequence of this is that each industry is deemed to hold the whole stock of its characteristic products and no other stocks.

The vectors a_1 and a_2 which enter into (24) are based on the experience of the last five or six years and cannot claim much precision. Their purpose is to divide a given sum available for competitive imports as a whole among the various categories. If a particular competitive import has shown itself in the past relatively insensitive to the state of the balance of payments, its component in a_1 will tend to be large while that in a_2 will tend to be small. Conversely, if it has shown itself relatively sensitive, its component in a_1 will tend to be small while that in a_2 will be large.

The diagonal matrix $\hat{a}_3$ which enters into (23) is based on post-war experience, allowances being made as far as possible for changes of a technical nature.

5. INDUSTRY BALANCES

The next set of balances I shall present relates to industries: their outputs and the inputs needed to make these outputs possible. Our computing programme provides not only the column sums of M, the domestic commodity outputs, shown in table 3, but also the row sums of M, the outputs of domestic industries. It also provides not only the row sums of X, the commodities entering into intermediate product, but also the column sums of X, the absorptions as intermediate product by the different industries of commodities supplied either from domestic production or as competitive imports. It also provides a vector of complementary imports into industries which completes the picture of intermediate inputs. The excess of an industry's output over its intermediate inputs represents its gross value added at market prices. This total for each industry must now be subdivided between: (i) indirect taxes less subsidies; (ii) depreciation; (iii) income from capital used in the industry; and (iv) income from employment in the industry. These calculations are made as follows.

The indirect taxes and subsidies are calculated at 1960 rates. In practice this means that customs duties are related to complementary imports and other indirect taxes and subsidies are related to industry outputs as they were in 1960.

There remains gross value added at factor cost. What we have done is first to divide this total for each industry into the amount accruing to capital and the amount accruing to labour and then to divide the amount accruing to capital between provisions for depreciation and income by relating depreciation to capital employed by the ratios appropriate to 1960.

The division of the net product of an industry between labour and capital is an undertaking which can only be carried out properly with the help of production functions. We have put a good deal of effort into the development of such functions but I shall not go into this subject at any length since it is fully treated by my colleague Graham Pyatt in [129]. I shall therefore limit myself to describing the method we have used so far [159] and shall not go on to later developments, as I did in the case of the consumption functions.

The method I shall now describe can be regarded as a natural development of social accounting. It applies, strictly, only to circumstances in which an industry is working at capacity but we have applied it in practice without checking the precise extent to which this condition is realized. In cases where it is clear that it is not realized we have made assumptions which we hope will save us from gross errors. Nevertheless, as presented here, these calculations must be considered tentative in the extreme: more than ever a basis for discussion rather than an expression of what we expect to happen.

Let us denote by Δy the increase in capacity net output in a particular industry in a particular year, let us denote by x the capacity added in the year as a result of new capacity coming into operation and let us denote by s the capacity subtracted through the scrapping of old plant. Then

$$\Delta y = x - s \tag{30}$$

Correspondingly, let Δl denote the increase in employment, let n denote the labour added to man the new plants and let r denote the reduction in employment due to plant retirements. Then

$$\Delta l = n - r \tag{31}$$

Let p denote the net output price, which is equal to the cost of labour and capital per unit of output, and let w denote the wage rate. Then if we multiply (30) by p and (31) by w and subtract, we obtain

$$p\Delta y - w\Delta l = (px - wn) - (ps - wr)$$
$$= (px - wn) \tag{32}$$

if plant is scrapped when it ceases to earn a return, that is when $ps = wr$.

Let us define the initial rate of return, r^* say, as the gross rate of return to capital embodied in new plant in the first year of its operation. Thus, denoting gross investment in new plant by v^*,

$$r^* = (px - wn)/v^*$$
$$= (p\Delta y - w\Delta l)/v^* \tag{33}$$

from (32) on the assumption that plant is scrapped when it ceases to earn a return. Equation (33) can be written either as

$$p\Delta y = w\Delta l + r^* v^* \tag{34}$$

which shows the value of the change in capacity net output, $p\Delta y$, divided into a part associated with labour, $w\Delta l$, and a part associated with capital, $r^* v^*$; or as

$$\Delta l = (p\Delta y - r^* v^*)/w \tag{35}$$

which shows the change in employment, Δl, as equal to the value of the change in capacity net output, $p\Delta y$, minus the part associated with capital, $r^* v^*$, all divided by the wage rate, w.

In making use of (35) to estimate the change in the labour force of each industry over the 1960's, it is necessary to estimate the variables $p\Delta y$, r^*, v^* and w. We took the increases in industrial net outputs, already calculated, to represent $p\Delta y$. We based our estimates of the initial rates of return, r^*, on the experience of the period 1948–1960, subject to a minimum rate of return of 5 per cent. We estimated investment in fixed assets over the decade, v^*, as five times the sum of the 1960 level and the 1970 level, the latter being obtained by adding together our estimates of industrial replacements and extensions. We estimated the industrial real wage rates over the decade, w, by increasing the 1960 wage rates in the different industries by one half of the increase in labour productivity required over the decade in industry as a whole. This last calculation required an estimate of the probable size of the labour force available to industry in 1970. By making the

TABLE 4 INDUSTRY BALANCES: BRITAIN 1970

(1960 £ million)

Industries	Revenues Total	Costs: Intermediate: domestic product and competitive imports	Intermediate: complementary imports	Net indirect taxes	Depreciation	Net income from property	Wages and salaries	Total	Annual growth rates 1960–70: Gross value added	Labour force	Labour productivity
1. Agriculture, forestry, fishing	1960	1016	0	−144	150	356	583	1960	2·4	−3·4	5·8
2. Coal mining	1011	296	3	10	53	−52	701	1011	2·1	−0·2	2·3
3. Mining and quarrying n.e.s.	208	99	0	4	12	26	67	208	2·8	−1·1	4·3
4. Food processing	2548	1529	459	−103	63	125	475	2548	2·2	0·5	1·7
5. Drink and tobacco	849	371	126	7	30	184	130	849	1·4	−1·6	3·0
6. Coke ovens, etc.	313	282	0	0	12	7	12	313	−0·9	−8·6	7·7
7. Mineral oil refining	896	247	583	16	25	498	18	896	4·2	4·3	8·5
8. Chemicals n.e.s.	3346	2192	132	31	114	153	379	3346	5·2	1·8	7·0
9. Iron and steel (m. r. and c.)	2679	1734	133	7	80	58	572	2679	4·4	2·7	1·7
10. Iron and steel (t. and t.)	532	375	0	2	10	24	88	532	3·6	0·5	3·1
11. Non-ferrous metals	1091	539	309	5	21	738	193	1091	5·0	1·7	1·3
12. Engineering and electrical goods	5821	3104	12	28	133	−41	1806	5821	4·1	−0·2	4·3
13. Shipbuilding and marine engineering	445	217	3	2	12	355	252	445	0·3	−0·2	0·5
14. Motors and cycles	3285	2338	5	13	49	124	525	3285	6·1	−0·3	6·4
15. Aircraft	1025	507	2	11	34	−25	346	1025	4·3	0·2	4·1
16. Railway locos. and rolling stock	231	133	4	1	5	182	113	231	2·0	1·4	0·6
17. Metals goods n.e.s.	2254	1393	95	10	43	33	531	2254	4·7	1·7	3·0
18. Textiles	2943	1751	349	30	75	178	706	2943	3·5	1·6	1·9
19. Leather, clothing, footwear	1798	1126	29	6	24	49	435	1798	6·0	1·2	4·8
20. Building materials	594	325	12	4	23	26	182	594	2·9	−0·2	3·1
21. Pottery and glass	256	118	0	2	8	8	102	256	1·5	1·7	3·2
22. Timber, furniture, etc.	809	365	136	6	18	100	184	809	3·7	−3·0	6·7
23. Paper, printing, publishing	2004	1025	123	11	63	407	376	2004	3·6	−4·6	8·2
24. Other manufacturing	1091	597	54	7	31	145	257	1091	6·3	0·8	5·5
25. Construction	4254	2424	56	19	72	17	1665	4254	3·6	1·0	2·6
26. Gas	484	252	0	14	54	−23	187	484	4·9	3·6	1·3
27. Electricity	1555	686	0	59	274	295	241	1555	7·8	5·0	2·8
28. Water	121	26	0	5	33	24	33	121	2·1	−2·8	4·9
29. Transport and communications	3950	1146	232	168	490	129	1786	3950	2·6	−0·3	2·9
30. Distributive traders	5475	939	0	275	233	659	3369	5475	4·3	1·9	2·4
31. Services n.e.s.	6537	1586	120	296	290	1024	3220	6537	4·8	0·5	4·3
Total or average	60363	28737	2977	802	2533	5780	19534	60363	4·0	0·5	3·5

Main source: computer run 20029. Components do not always add up to totals because of rounding-off errors.

necessary adjustments to the official estimates of the probable supply of labour for all purposes in 1970, we reached the conclusion that the industrial labour force might be expected to grow by about 5 per cent over the decade.

By following these steps, we obtained an estimate for Δl in each industry over the 1960's. By adding these changes to the corresponding labour forces in 1960, we obtained estimates for 1970. These calculations suggested a total industrial demand for labour 4 per cent in excess of the estimated supply. At this stage we did not attempt to resolve this discrepancy but simply scaled up the initial rates of return, r^*, by a constant factor and thus adjusted the total demand for labour to the estimated supply.

Our final step was to multiply the labour force in each industry in 1970 by the corresponding real wage rate and thus obtain estimates of the 1970 wage bills expressed at 1960 values.

By subtracting the wage bill of an industry from the corresponding estimate of gross value added at factor cost, we obtain an estimate of the gross return to capital. These estimates are composed of property income plus provisions for depreciation. The latter estimates were obtained by applying the 1960 depreciation rates to the implicit capital stocks of 1970.

The results of these calculations are brought together in table 4, which also shows the annual growth rates over the 1960's in gross value added, the labour force and labour productivity for each industry. As I have said, the subdivision of the estimates of gross value added at factor cost is highly uncertain. The reason for this is not simply that we have used relationships of the form of (35) in place of production functions to estimate the changes in the labour forces of the different industries; the figures in table 4 are obtained after only one round of calculations, and so make no allowance for changes in relative prices after 1960. Even with the present method of allocating labour, this could be done, but the steps involved have not yet been programmed and so, at present, the model is incomplete in this respect. Yet, we can see from the table the way in which allowances for relative price changes would influence the figures. On the assumption that demands are not very sensitive to price and that the relative prices of an industry's inputs are not, on average, very different from what they were in 1960, we should expect the relative price of the product of an industry to fall if its labour productivity is rising above average, and to rise if its labour productivity is rising below average. Thus we should expect the relative prices of chemical products and cars to fall and of coal and ships to rise. We should also expect relative wage rates, compared with 1960, to move in favour of labour with highly growing productivity. When we are able to allow for these factors we may expect to see a less extreme picture emerging from the calculations. One of our next tasks is to programme the model to calculate and respond to 1970 shadow prices. In the meantime the figures in table 4, though exaggerated, may be useful as an indication of tendencies and also as a reminder of the

TABLE 5 INDUSTRIAL FIXED INVESTMENT BALANCES: BRITAIN 1970

(1960 £ million)

Industries	Industrial replacements	Industrial extensions	Gross investment demands	Depreciation	Net investment	Gross investment finance
1. Agriculture, forestry, fishing	97	69	166	150	16	166
2. Coal mining	19	177	196	53	143	196
3. Mining and quarrying n.e.s.	7	12	19	12	7	19
4. Food processing	20	88	108	63	45	108
5. Drink and tobacco	16	27	43	30	13	43
6. Coke ovens, etc.	3	25	28	12	16	28
7. Mineral oil refining	1	56	57	25	32	57
8. Chemicals n.e.s.	29	193	222	114	108	222
9. Iron and steel (m. r. and c.)	24	36	60	80	−20	60
10. Iron and steel (t. and t.)	3	4	7	10	−3	7
11. Non-ferrous metals	6	45	51	21	30	51
12. Engineering and electrical goods	47	214	261	133	128	261
13. Shipbuilding and marine engineering	3	15	18	12	6	18
14. Motors and cycles	14	90	104	49	55	104
15. Aircraft	9	45	54	34	20	54
16. Railway locos. and rolling stock	5	4	9	5	4	9
17. Metal goods n.e.s.	18	83	101	43	58	101
18. Textiles	57	40	97	75	22	97
19. Leather, clothing, footwear	11	67	78	24	54	78
20. Building materials	8	29	37	23	14	37
21. Pottery and glass	2	13	15	8	7	15
22. Timber, furniture, etc.	7	34	41	18	23	41
23. Paper, printing, publishing	29	95	124	63	61	124
24. Other manufacturing	7	76	83	31	52	83
25. Construction	35	104	139	72	67	139
26. Gas	35	53	88	54	34	88
27. Electricity	146	450	596	274	322	596
28. Water	30	24	54	33	21	54
29. Transport and communications	377	334	711	490	221	711
30. Distributive trades	106	361	467	233	234	467
31. Services n.e.s.	106	507	613	290	323	613
Total	1277	3368	4645	2533	2112	4645

importance of allowing for changes in relative prices even if demands for commodities are not very sensitive to these changes.

A corresponding table for 1960 can be formed from table 2 of [31].

6. BALANCES FOR INDUSTRIAL FIXED ASSETS

In table 2, I gave estimates of industrial replacements grouped according to the industry whose product would be needed to meet these demands. This information was obtained from a sub-matrix which, in the complete system, would occupy a position at the intersection of row 1 and column 8 of table 1. If the elements of this matrix are added up by columns rather than by rows, and if purchase taxes on vehicles are added, we obtain a vector of industrial replacements classified by the using industry. From our computing programme we obtain a similar vector for industrial extensions. By adding these two vectors together we obtain a vector of gross investment in fixed assets by the different industries in 1970.

In table 4, I have given a vector of provisions for depreciation by the different industries in 1970. By subtracting the elements of this vector from the corresponding elements of the vector of gross investments in fixed assets, we obtain a vector of net investments. This set of balances is shown in table 5. A corresponding table for 1960 can be formed from tables 8 and 9 of [31].

7. FOREIGN TRADE BALANCES

The first questions that come to the mind under this heading are whether the external account balances and, if so, how this balance is brought about by the model. We have already seen from table 1 that a balance is achieved; the way in which it comes about is as follows.

The favourable balance of payments of £200 million, shown at the intersection of row 15 and column 14 of table 1, is the result of setting values on net income from foreign investments received by Britain, £250 million, net current remittances abroad made by Britain, £100 million, and Britain's balance of trade, $\beta = £50$ million: $£(250 + 50 - 100)$ million $= £200$ million. The first two of these figures are in line with recent experience, but we have taken no steps at this stage to examine future prospects here in detail. The figure of £50 million for β is fixed arbitrarily and reflects a desire to achieve a small excess of exports over imports. In the version of the model I have described, exports are estimated as part of exogenous final demand, and this applies not only to the exports of industrial commodities, £7151 million, but also to exports in the form of expenditure by foreign visitors to Britain, £320 million. On the side of imports, exogenous estimates are made of complementary imports into private consumption, £780 million, and into government purposes, £194 million. From these figures we obtain £(7151 + 320 − 50 − 780 − 194) million = £6447 million as the sum available for all other imports, that is to say competitive imports of commodities and complementary imports into industries including the small item of additions to the stock of these imports. These estimates are obtained from (23), (25) and (28) of section 4 above, together with an equation for additions to stocks of complementary imports which, for simplicity, I left out of my description of the way in which the commodity balances are achieved. In this example we obtained for the three import totals £(3456 + 2977 + 14) million = £6447 million, as required.

We have already seen in section 3 above that, for present purposes, we have assumed a rise of 51 per cent between 1960 and 1970 in the total of British exports of commodities. We can see from table 1 that the corresponding increases for competitive imports and for complementary imports into industries are 40 and 36 per cent. Since the assumptions used in this particular run of the computing programme are designed to bring about a large improvement in the balance of payments, an improvement of £544 million compared with 1960, it is to be expected that exports must rise faster than imports over the 1960's. But since the brunt of the balance of payments adjustment is borne, in this version of the model, by competitive imports, it is interesting to see that in total they are shown as increasing by rather more than complementary imports into industries.

In [30] two variants of the foreign trade mechanism described above are set out. I mention them here because they show that the model can easily be adapted either to allow the different competitive imports to be linear functions of all direct and indirect demands generated in the economy or to allow exports no longer to be estimated exogenously but to be determined by import requirements and the balance of payments constraint.

The second question, to which I shall now turn, relates to the contribution made by each branch of commodity production to the balance of trade. It seems reasonable to take the country's exports of each commodity and subtract the corresponding figure for competitive imports; the question remains, what should be subtracted further in respect of complementary imports? This can hardly be the complementary imports needed directly to produce the commodity in question, since this measure would overlook two things. In the first place, indirect as well as direct complementary imports are needed and so the simple measure would not include all the complementary imports required; and, in the second place, a part of the production of any commodity goes to meet intermediate rather than final demand and so is used indirectly in some other form of production. Accordingly, the figure to be subtracted in respect of complementary imports should be defined as the direct and indirect complementary imports needed by a branch of commodity production to meet its production for final purposes only.

The algebra of this measure is as follows. The flow equation for commodities, (16), can be matched by a similar equation for commodity prices, the elements of a vector p say. This equation takes the form

$$p = A'p + M^{-1}(t+y+n)$$
$$= (I-A')^{-1} M^{-1}(t+y+n) \tag{36}$$

where A' denotes the transpose of A, M^{-1} denotes the inverse of M, and t, y and n denote respectively vectors of net indirect taxes, payments for labour and capital (including depreciation) and complementary imports charged to the production accounts of industries. The effect of premultiplying these charges by M^{-1} is to convert them into vectors of the corresponding charges per unit of commodity production. Consequently (36) states simply that the price of any commodity can be built up from the costs of intermediate product (including complementary imports), net indirect taxes, and charges for labour and capital, all expressed per unit of output.

If we premultiply (36) by $\hat{f}$, a diagonal matrix of final demands for commodities, we obtain a vector of the contribution which each branch of commodity production makes to final product. We can decompose this vector as follows:

$$\hat{f}p = \hat{f}(I-A')^{-1} M^{-1}(t+y) + \hat{f}(I-A')^{-1} M^{-1}n \tag{37}$$

The second term on the right-hand side of (37) is a vector whose elements represent the complementary imports needed directly and indirectly to enable a branch of commodity production to produce its contribution to final product. This term can be decomposed in two ways:

$$\hat{f}(I-A')^{-1} M^{-1}n = \hat{f}M^{-1}n + \hat{f}A'(I-A')^{-1} M^{-1}n$$
$$= (\hat{h}+\hat{v})(I-A')^{-1} M^{-1}n + (\hat{x}-\hat{m})(I-A')^{-1} M^{-1}n \tag{38}$$

In the first row of (38), the requirements for complementary imports are divided between a direct and an indirect component; and in the second row they are divided between a domestic and a foreign component. This foreign component consists of two parts: direct and indirect requirements in respect of exports less the savings of complementary imports due to the fact that part of the supply of commodities comes from abroad in the form of competitive imports.

The results of these calculations are set out for 1970 in table 6. In this case figures for 1960 are not readily available, though the information needed to compile them is given in [31, 32].

The sum of the balances in each of these tables is the excess of exports over competitive imports plus complementary imports absorbed in industrial production. We can see this algebraically if we premultiply (37) by i' and remember from (16) that $f'(I-A')^{-1}=q'$. In this way we obtain

$$f'p = q'M^{-1}(t+y) + q'M^{-1}n \tag{39}$$

19

TABLE 6 FOREIGN TRADE BALANCES: BRITAIN 1970
(1960 £ million)

Characteristic products of	Exports Total	Imports Competitive	Imports Direct complementary into final demand	Imports Indirect complementary into final demand	Imports Total complementary	Imports Total	Balance Exports minus imports	Alternative treatment of complementary imports Into net exports	Alternative treatment of complementary imports Into other final demand
1. Agriculture, forestry, fishing	107	335	0	144	144	479	−372	−20	165
2. Coal mining	69	1	1	9	10	11	58	2	8
3. Mining and quarrying n.e.s.	33	29	0	1	1	30	3	0	1
4. Food processing	146	531	334	137	471	1002	−856	−110	581
5. Drink and tobacco	167	61	124	41	164	225	−58	21	143
6. Coke ovens, etc.	9	2	0	3	3	5	4		3
7. Mineral oil refining	50	106	262	27	288	394	−344	−41	329
8. Chemicals n.e.s.	818	347	33	84	117	464	354	51	65
9. Iron and steel (m. r. and c.)	240	26	11	15	26	52	188	26	0
10. Iron and steel (t. and t.)	116	1	0	10	10	11	105	10	0
11. Non-ferrous metals	182	60	29	10	39	99	83	50	−11
12. Engineering and electrical goods	1316	393	−9	243	234	627	689	56	179
13. Shipbuilding and marine engineering	81	21	2	16	18	39	42	3	14
14. Motors and cycles	1173	143	3	213	216	359	814	83	133
15. Aircraft	259	60	1	14	15	75	184	6	9
16. Railway locos, and rolling stock	3	0	1	10	12	12	−9	0	12
17. Metal goods, n.e.s.	207	81	14	36	51	132	75	16	34
18. Textiles	284	238	45	42	87	325	−41	−8	79
19. Leather, clothing, footwear	85	127	26	178	205	332	−247	−7	212
20. Building materials	25	16	1	2	3	19	6	1	2
21. Pottery and glass	55	32	0	6	6	38	17	1	5
22. Timber, furniture, etc.	19	106	31	13	44	150	−131	−21	66
23. Paper, printing, publishing	166	160	21	17	38	198	−32	−1	39
24. Other manufacturing	134	77	18	26	44	121	13	6	37
25. Construction	10	4	43	205	247	247	−237	−1	247
26. Gas	4	17	0	11	11	28	−24	−1	12
27. Electricity	6	0	0	39	39	39	−33	0	38
28. Water	12	0	0	2	2	2	10	0	2
29. Transport and communications	836	28	122	78	200	229	607	77	123
30. Distributive trades	16	8	−1	115	114	122	−106	0	114
31. Services n.e.s.	523	450	56	69	124	574	−51	3	121
Total	7151	3456	1168	1809	2977	6433	718	215	2762

Main source: computer run 20029. Components do not always add up to totals because of rounding-off errors.

For present purposes exports have been defined as net exports of commodities and the second term on the right-hand side of (39) is simply the vector product of domestic commodity outputs and complementary imports per unit of these outputs. This product is equal to the total of complementary imports into industries, since $q'M^{-1}=i'$.

8. FINANCIAL BALANCES

As our model stands at present it does not place any financial constraint on what the economy can achieve. If we go back to table 1, the entry at the intersection of row 14 and column 6, £4680 million, is the community's saving of the year and is, of course, equal to the community's net investment at home and abroad, the components of which appear in column 14. This is necessarily the case, since the model satisfies a complete set of accounting constraints. But we must still ask whether it seems likely that the various institutional sectors of the economy would wish to save in the aggregate the amount required to finance net investment if their incomes were as the model suggests that they will be.

In order to answer this question, we must establish saving functions and then try to discover whether the supply of saving would equal the demand under the conditions assumed.

We have approached this problem by constructing saving functions for persons and companies. This development was foreshadowed in [167] and set out in greater detail in [XV]. These equations, in which saving depends on income, wealth, the preceding year's income and the preceding year's expenditure, make it possible to calculate equilibrium saving ratios for a steady rate of growth of income. Accordingly, our first answer to the question of financial balances is based on the equilibrium saving ratios of persons and companies calculated on the assumption that in 1970 their disposable incomes are growing at 4 per cent in real terms and that transient effects can be ignored. Let us now see how these calculations work out.

In setting out this train of ideas, I shall concentrate on the basic form of the saving relationship and ignore a number of complications which can easily be introduced. As is shown in [167, XV], this basic form can be expressed as

$$\sigma = [1-\beta_1\lambda-\beta_2(1-\lambda)]\mu-\alpha_1\lambda\omega+\beta_2(1-\lambda)E^{-1}\mu-(1-\lambda)E^{-1}\epsilon \quad (40)$$

In this equation, the variables σ, ω, μ and ϵ relate respectively to saving, wealth (at the beginning of the year), disposable income and expenditure. These variables are connected by two identities: $\sigma \equiv \mu-\epsilon$, saving is equal to disposable income less expenditure; and $\sigma \equiv \dot{\omega}$, saving is equal to the rate of change of wealth. All these variables are measured at constant consumer prices.

19*

The parameters in (40), $\alpha_1, \beta_1, \beta_2$ and λ have the following meanings. The first two, α_1 and β_1, represent the marginal propensities to consume out of the permanent components of wealth and income. The third, β_2, represents the marginal propensity to consume out of the transient component of income. And, finally, λ represents the proportion of the excess of this year's income (or wealth or expenditure) over last year's permanent component, which is considered to be permanent. The operator E^{-1} indicates the preceding year's value of the variable to which it is applied: $E^{-1}\mu$ indicates last year's disposable income.

The permanent and transient components of income and wealth cannot be observed and (40) is expressed in a form which connects only observable variables. If we denote the permanent and transient components of income by μ_1 and μ_2, so that $\mu \equiv \mu_1 + \mu_2$, we can work out the equilibrium values of the saving ratio as follows.

First, as shown in [167, XV], we can express μ_1 in terms of μ by the relationship

$$\mu_1 = \lambda \sum_{\theta=0}^{\infty} (1-\lambda)^\theta E^{-\theta} \mu \tag{41}$$

where $E^{-\theta}\mu$ denotes disposable income θ years ago. If μ grows exponentially at a rate ρ, that is if

$$\mu = \bar{\mu} e^{\rho t} \tag{42}$$

where $\bar{\mu}$ denotes the value of μ at time $t=0$, then it follows, by combining (41) and (42), that

$$\frac{\mu_1}{\mu} = \frac{\lambda}{1-(1-\lambda)e^{-\rho}}$$

$$= \zeta \tag{43}$$

say. The variables ω and ω_1 are connected in exactly the same way. For steady growth in μ, we can now write

$$\dot{\omega} \equiv \sigma$$

$$\equiv \mu - \epsilon$$

$$= [1 - \beta_1 \zeta - \beta_2(1-\zeta)]\mu - \alpha_1 \zeta \omega$$

$$\equiv \phi\mu - \psi\omega \tag{44}$$

The solution of this differential equation is

$$\omega = \kappa e^{-\psi t} + \frac{\phi\mu}{\psi + \rho} \tag{45}$$

If we differentiate this equation with respect to time and divide by μ, we obtain

$$\frac{\sigma}{\mu} = -\frac{\kappa\psi e^{-\psi t}}{\mu} + \frac{\phi\rho}{\psi + \rho} \tag{46}$$

as an expression for the saving ratio under conditions of steady growth in income. It is made up of two terms representing the transient and the steady-state components. The transient component tends to zero with time since ψ is positive; and, as can be seen by putting $t = 0$ in (45),

$$\kappa = \bar{\omega} - \frac{\phi\bar{\mu}}{\psi + \rho} \tag{47}$$

will always be zero if the initial level of wealth at the beginning of the steady state happens to be equal to the equilibrium level associated with the rate of growth of income that we are considering.

In discussing 1970, we can reasonably ignore the transient term in (46) and concentrate on the steady-state term. In [XV] values of the equilibrium saving ratio, calculated in this way by applying (40) to post-war data, are given for persons and companies. Let us consider values of ρ equal to 0, 2 per cent, 4 per cent and 6 per cent: then for persons the corresponding values of σ/μ work out to 0, 6·1 per cent, 10·2 per cent and 13·2 per cent; and for companies these ratios are 0, 23·8 per cent, 38·7 per cent and 48·7 per cent.

Since 1948, the personal saving ratio has pursued an irregular upward course. It was negative until 1952, reached a peak of 7·5 per cent in 1961, fell back to under 6 per cent in 1962 and is now slowly recovering. So the value of 10·2 per cent calculated for a steady 4 per cent growth is higher than anything in recent experience.

The history of company saving is quite different. The saving ratio for companies has pursued an irregular downward course with a peak of 72·4 per cent in 1951 and a low of 48·6 per cent in 1962. So the value of 38·7 per cent calculated for a steady 4 per cent growth is lower than anything in recent experience.

We are now in a position to attempt a trial financial balance for the economy in 1970. Since, at present, our model does not allocate income to the different institutional sectors, I have based my calculations on the allocations as they actually were in the three-year period 1960–1962. I have calculated the disposable income of persons and of companies in these three years and applied to the two totals the equilibrium saving ratios appropriate to 4 per cent growth. From these totals I have deducted company stock appreciation as it actually was in the three-year period, since this item is included in saving derived from the company saving function but not in the concept of saving used in the model. The total of private saving thus obtained is £8556 million compared with an actually realized figure of £8191 million. Over the same three years, total income, defined as the national income plus net indirect taxes, amounted to £74198 million. The ratio of hypothetical private saving to income is thus 8556/74198 = 11·5 per cent.

If we now turn to table 1 we can make the corresponding calculation of the demand for saving. At the intersection of row 14 and column 6, we see a figure of £4680 million which represents total saving on the definitions employed in the model. From this figure we must subtract £852 million in respect of net investment in consumers' durables (excluding dwellings) which in the personal saving function have been treated as part of consumers' expenditure. Since, in table 1, total income is £34305 million, we obtain a demand ratio for saving of 3828/34305 = 11·2 per cent.

Thus on this trial balance, the private sector would provide a little more than the whole of the saving needed to meet the demand in 1970. But the relative importance of personal and company saving would completely change in comparison with the position in 1960–62. Thus, in the hypothetical calculation, personal saving is 67 per cent of private saving whereas over the years 1960–62 it was actually 43 per cent.

9. THE END OF THE BEGINNING

This is as far as I can go at present in providing a set of economic balances for 1970. Apart from all the shortcomings of the work itself, the picture is obviously not complete; I have said nothing, for example, about labour skills, the distribution of incomes or the regional balance of the economy. In a sense, however, this picture marks the end of the beginning of our work; we have assembled a body of data, a set of econometric relationships and a computing programme which enable calculations of this kind to be made. From now on we have a working model to help us in the quantitative study of economic interdependence. What we must now do is to study its implications, refine and extend it.

10. THE NEXT STAGE

I shall now give a brief sketch of the next stages of our work: not the stages that we have vaguely at the back of our minds for attention in the future but those on which we are now actively engaged.

First, there is the improvement of the basic data on which the model is built. I have explained the steps we have taken with regard to private consumption, and my colleague Graham Pyatt will speak of the work which is relevant to industrial production. The main new development we are planning in this area is the establishment of a small group which will compile a regular series of social accounting matrices. We expect to learn much from the gradual changes shown in these matrices and to be able, with their help, to test the present performance of the relationships used in our model.

Second, in carrying out this factual work, we shall extend our social accounting matrix to embody a treatment of financial capital transactions, or flows-of-funds, and also a set of balance sheets for the main sectors of the economy.

Third, the work of compiling economic information so as to form a coherent system must start with the analysis of official statistics, since this body of information has a wide coverage and, on the whole, consistent definitions and classifications. But the picture we can reach is only approximate and, in some respects, a little out of date. We attach great importance to the co-operation of practical experts in various branches of the economy whose special knowledge alone makes it possible to add substantially greater realism to the initial estimates. Obviously, this kind of assistance becomes even more important as we move into the future and have to consider changes from a more or less known present.

Fourth, having discussed the factual basis of our model we come next to the question of relationships. Here, the first thing to recognize is that, as it stands, the model is incomplete. It presents a consistent picture of 1970 but it does not go into the steps that would have to be taken in the 1960's so that the 1970 picture could be realized. This second part of the model, as we see it at present, has been described in [34, VIII] and a computing programme for it has been outlined in [143]. The essential idea is that the equipment needed to provide the capacity for output in 1970 must have been installed beforehand and we ought to check that this can in fact be done while preserving a reasonably smooth growth in consumption from the past, through the transitional period from now to 1970, and after 1970. The object of our second model, therefore, is to maximize consumption through the transitional period subject to fixed initial stocks and minimum terminal stocks and subject also to a constraint on the minimum level of consumption acceptable at any one time.

Fifth, in addition to the refinement of existing relationships in the model, which I have mentioned at various points in the paper, we are also starting to work on new relationships which have not so far been embodied in it. I mentioned the analysis of private saving in section 8 above. Another problem on which we have also made some progress is the changing pattern of labour skills which accompanies technical progress [27, 159]. So far we have concentrated on the demand for skills but we intend also to examine the supply position as determined by arrangements for education and training.

Sixth, as all these complications are introduced, the model itself grows in complexity but not in the detail in which it treats elements already included, such as production or consumption. As I have already mentioned, we are planning to experiment with a sub-model for the energy industries since we think it better, when a model has reached a certain size, to work in terms of a set of linked models rather than in terms of a single model which is like the original one only larger still. With a system of models, each model has an exogenous and an endogenous part. If the models form a hierarchy, the exogenous part of the sub-models will be supplied with information from the main model and, in turn, at the next stage in the calculations will modify the exogenous input into the main model. We are at present experimenting with the programming problems of such model-systems. If they turn out to be workable, they offer two great advantages: the main model can be kept to a manageable size; and each of the sub-models can be built and operated by people with a greater knowledge of their subject matter, whether this be the energy industries or the educational system, than can be expected of the group working on the main model.

11. CONCLUSION

This, then, is the stage we have reached and an indication of the direction we are going in. At present our model is a working prototype, not the full scale version we hope eventually to build. It is designed to show the economy working as an interdependent system and thus to enable us to see in quantitative terms the apparent consequences of trying to reach certain objectives and the difficulties that such attempts might be expected to lead to. We believe that in economic policy-making too little attention is paid to interdependence, a phenomenon easy to observe but difficult to handle. By studying the changing interdependencies of our actual economic system, we hope to increase the flow of information in it and so contribute indirectly to better economic decisions.

A LIST OF WORKS CITED

1. AITCHISON, J., and J. A. C. BROWN. A synthesis of Engel curve theory. *The Review of Economic Studies*, vol. XXII, 1954–55, pp. 35–46.
2. AITCHISON, J., and J. A. C. BROWN. *The Lognormal Distribution*. Cambridge University Press, 1957.
3. AITCHISON, J., and S. D. SILVEY. Maximum-likelihood estimation of parameters subject to constraints. *The Annals of Mathematical Statistics*, vol. 29, no. 3, 1958, pp. 813–28.
4. ALLEN, R. G. D. *Mathematical Analysis for Economists*. Macmillan, London, 1938.
5. ALLEN, R. G. D. *Mathematical Economics*. Macmillan, London, 1956.
6. ALLEN, R. G. D. The structure of macro-economic models. *The Economic Journal*, vol. LXX, no. 277, 1960, pp. 38–56.
7. ARROW, Kenneth, Hollis B. CHENERY, Bagicha MINHAS and Robert M. SOLOW. Capital-labor substitution and economic efficiency. *The Review of Economics and Statistics*, vol. XLIII, no. 3, 1961, pp. 225–50.
8. ARTLE, Roland. *Studies in the Structure of the Stockholm Economy*. Business Research Institute, Stockholm School of Economics, Stockholm, 1959.
9. BAILEY, Norman T. J. *The Mathematical Theory of Epidemics*. Griffin, London, 1957.
10. BAIN, A. D. The growth of demand for new commodities. *Journal of the Royal Statistical Society, Series A*, vol. 126, pt. 2, 1963, pp. 285–99.
11. BAIN, A. D. *The Growth of Television Ownership in the United Kingdom*. Cambridge University Press, 1964.
12. BARONE, Enrico. Il ministro della produzione nello stato collettivista. *Giornale degli Economisti*, vol. 37, 1908, pp. 267–93. An English translation appeared as appendix A, pp. 245–90, to *Collectivist Economic Planning*, Routledge and Kegan Paul, London, 1935.
13. BARTLETT, M. S. *Stochastic Population Models*. Methuen, London; John Wiley, New York; 1960.
14. BATCHELOR, James H. *Operations Research: An Annotated Bibliography*. Saint Louis University Press, 1959.
15. BAUCHET, Pierre. *Les Tableaux Economiques: Analyse de la Région Lorraine*. Genin, Paris, 1955.
16. BELLMAN, Richard. *Dynamic Programming*. Princeton University Press, 1957.
17. BELLMAN, Richard, and Stuart E. DREYFUS. *Applied Dynamic Programming*. Princeton University Press, 1962.
18. BENTZEL, Ragnar, and Herman WOLD. On statistical demand analysis from the viewpoint of simultaneous equations. *Skandinavisk Aktuarietidskrift*, vol. XXIX, no. 1/2, 1946, pp. 95–114.
19. BERGSTROM, A. R. An econometric study of supply and demand for New Zealand's exports. *Econometrica*, vol. 23, no. 3, 1955, pp. 258–76.
20. BOSE, R. C. On the exact distribution and moment coefficients of D^2-statistics. *Sankhyā*, vol. 2, pt. 2, 1936, pp. 143–54.
21. BOSE, R. C. A note on the distribution of differences in mean values of two samples drawn from two multivariate normally distributed populations, and the definition of the D^2-statistic. *Sankhyā*, vol. 2, pt. 4, 1936, pp. 379–84.

22. BOSE, R. C., and S. N. ROY. The distribution of Studentised D^2-statistic. *Sankhyā*, vol. 4, pt. 1, 1938, pp. 19–38.

23. BOSE, S. N. On the complete moment-coefficients of the D^2-statistic. *Sankhyā*, vol. 2, pt. 4, 1936, pp. 385–96.

24. BOSE, S. N. On the moment-coefficients of the D^2-statistics and certain integral and differential equations connected with the multivariate normal population. *Sankhyā*, vol. 3, pt. 3, 1937, pp. 105–24.

25. BRIGGS, F. E. A. On problems of estimation in Leontief models. *Econometrica*, vol. 25, no. 3, 1957, pp. 444–55.

26. BRIGGS, F. E. A. The estimation of regression equations when the independent variables are otherwise related to the dependent variables. *Metroeconomica*, vol. XII, no. 2–3, 1960, pp. 39–57.

27. BROWN, Alan, Colin LEICESTER and Graham PYATT. Output, manpower and industrial skills in the United Kingdom. In *The Residual Factor and Economic Growth*. O.E.C.D., Paris, 1964.

28. BROWN, J. A. C., and L. J. SLATER. *On the RAS technique*. Mimeographed, 1961.

29. BUSH, Robert R., and Frederick MOSTELLER. *Stochastic Models of Learning*. John Wiley, New York; Chapman and Hall, London, 1955.

30. CAMBRIDGE, DEPARTMENT OF APPLIED ECONOMICS. *A Computable Model of Economic Growth*. No. 1 in *A Programme for Growth*. Chapman and Hall, London, 1962.

31. CAMBRIDGE, DEPARTMENT OF APPLIED ECONOMICS. *A Social Accounting Matrix for 1960*. No. 2 in *A Programme for Growth*. Chapman and Hall, London, 1962.

32. CAMBRIDGE, DEPARTMENT OF APPLIED ECONOMICS. *Input–Output Relationships, 1954–1966*. No. 3 in *A Programme for Growth*. Chapman and Hall, London, 1963.

33. CAMBRIDGE, DEPARTMENT OF APPLIED ECONOMICS. *Capital, Output and Employment, 1948 to 1960*. No. 4 in *A Programme for Growth*. Chapman and Hall, London, 1964.

34. CAMBRIDGE, DEPARTMENT OF APPLIED ECONOMICS. *The Model in Its Environment: a Progress Report*. No. 5 in *A Programme for Growth*. Chapman and Hall, London, 1964.

35. CAMBRIDGE, DEPARTMENT OF APPLIED ECONOMICS. *Exploring 1970: Some Numerical Results*. No. 6 in *A Programme for Growth*. Chapman and Hall, London, 1965.

36. CAMBRIDGE, DEPARTMENT OF APPLIED ECONOMICS. *An International Survey of Planning Models*. To be published in *A Programme for Growth*. Chapman and Hall, London.

37. CANADA, DOMINION BUREAU OF STATISTICS. *National Accounts Income and Expenditure 1926–1956*. Dominion Bureau of Statistics, Ottawa, 1958.

38. CAO-PINNA, Vera. Principali caratteristiche strutturali di due economie mediterrenee: Spagna e Italia. *Economia Internazionale*, vol. XI, no. 2, 1958, pp. 259–311.

39. CARNAP, R. On inductive logic. *Philosophy of Science*, vol. 12, 1945, pp. 72 ff.

40. CHENERY, Hollis B. Inter-regional and international input–output analysis. In *The Structural Interdependence of the Economy*. John Wiley, New York; Giuffrè, Milano; 1955; pp. 341–56.

41. CHENERY, Hollis B., and Paul G. CLARK. *Interindustry Economics*. John Wiley, New York, 1959.

42. CHENERY, Hollis B., Paul G. CLARK and Vera CAO-PINNA. *The Structure and Growth of the Italian Economy*. U.S. Mutual Security Agency, Rome, 1953.

43. CHENERY, Hollis B., and Tsunehiko WATANABE. International comparisons of the structure of production. *Econometrica*, vol. 26, no. 4, 1958, pp. 487–521.

44. COWLES COMMISSION FOR RESEARCH IN ECONOMICS. *Studies in Econometric Method*. John Wiley, New York; Chapman and Hall, London; 1953.

45. D'ALEMBERT, Jean le Rond. Croix ou pile. *Encyclopédie ou Dictionnaire Raisonné*, 1754.

46. DEANE, Phyllis. Regional variations in United Kingdom incomes from employment, 1948. *Journal of the Royal Statistical Society, Series A*, vol. CXVI, pt. II, 1953, pp. 123–39.

47. DEMING, W. Edwards. *Statistical Adjustment of Data*. John Wiley, New York, 1943.

48. DERWA, Léon. Une nouvelle methode d'analyse de la structure économique. *Revue du Conseil Économique Wallon*, no. 28, 1957, pp. 16–42.

49. DERWA, Léon. Technique d'input–output et programmation linéaire. *Revue du Conseil Économique Wallon*, no. 34, 1958, pp. 35–58.

50. DORFMAN, Robert, Paul A. SAMUELSON and Robert M. SOLOW. *Linear Programming and Economic Analysis*. McGraw-Hill, New York, 1958.

51. DURBIN, J. *The adjustment of observations subject to linear constraints*. Mimeographed.

52. DURBIN, J. A note on regression when there is extraneous information about one of the coefficients. *Journal of the American Statistical Association*, vol. 48, no. 264, 1953, pp. 799–808.

53. DURBIN, J. *Maximum-likelihood estimation of the parameters of a system of simultaneous regression equations*. Joint European Conference of the Institute of Mathematical Statistics and the Econometric Society, 1963. Mimeographed.

54. FINNEY, D. J. *Probit Analysis: A Statistical Treatment of the Sigmoid Response Curve*. Cambridge University Press, 1947; second edition 1952.

55. FISHER, M. R. A sector model – the poultry industry of the U.S.A. *Econometrica*, vol. 26, no. 1, 1958, pp. 37–66.

56. FISK, P. R. Maximum likelihood estimation of Törnqvist demand equations. *The Review of Economic Studies*, vol. XXVI, 1958–59, pp. 33–50.

57. FORSYTH, F. G. The relationship between family size and family expenditure. *Journal of the Royal Statistical Society, Series A*, vol. 123, pt. 4, 1960, pp. 367–97.

58. FOX, Karl A. *Econometric Analysis for Public Policy*. Iowa State College Press, 1958.

59. GALE, David. *The Theory of Linear Economic Models*. McGraw-Hill, New York, 1960.

60. GEARY, R. C. A note on 'A constant-utility index of the cost of living'. *The Review of Economic Studies*, vol. XVIII (1), no. 45, 1949–50, pp. 65–6.

61. GEARY, R. C. A note on the comparison of exchange rates and purchasing power between countries. *Journal of the Royal Statistical Society, Series A*, vol. 121, pt. 1, 1958, pp. 97–9.

62. GILBERT, Milton, and associates. *Comparative National Products and Price Levels*. O.E.E.C., Paris, 1958.

63. GLASS, D. V. (editor). *Social Mobility in Britain*. Routledge and Kegan Paul, London, 1954.

64. GOLDBERG, Samuel. *Introduction to Difference Equations*. John Wiley, New York, 1958.
65. GOODWIN, Richard M. Iteration, automatic computers and economic dynamics. *Metroeconomica*, vol. III, no. 1, 1951, pp. 1–7.
66. GOODWIN, R. M. The nonlinear accelerator and the persistence of business cycles. *Econometrica*, vol. 19, no. 1, 1951, pp. 1–17.
67. HAAVELMO, Trygve. The statistical implications of a system of simultaneous equations. *Econometrica*, vol. 11, no. 1, 1943, pp. 1–12.
68. HANNA, Frank A. *State Income Differentials 1919–1954*. Duke University Press, Durham, N.C., 1959.
69. HEADY, Earl O., and Wilfred CANDLER. *Linear Programming Methods*. Iowa State College Press, 1958.
70. HICKS, J. R. *Value and Capital*. Clarendon Press, Oxford, 1939.
71. HITCHCOCK, Frank L. The distribution of a product from several sources to numerous localities. *Journal of Mathematics and Physics*, vol. 20, 1941, pp. 224–30.
72. HOTELLING, Harold. Analysis of a complex of statistical variables into principal components. *Journal of Educational Psychology*, vol. XXIV, 1933, pp. 417–41 and 498–520.
73. HURWICZ, Leonid. On the structural form of interdependent systems. In *Logic, Methodology and Philosophy of Science*. Stanford University Press, 1962.
74. ISARD, Walter. Inter-regional and regional input–output analysis: a model of a space-economy. *The Review of Economics and Statistics*, vol. XXXIII, no. 4, 1951, pp. 318–28.
75. ISARD, Walter, Eugene W. SCHOOLER and Thomas VIETORISZ. *Industrial Complex Analysis and Regional Development*. The Technology Press, Massachusetts Institute of Technology; John Wiley, New York; Chapman and Hall, London; 1959.
76. JAPAN, KANSAI ECONOMIC FEDERATION. *Inter-regional Input–Output Table for the Kinki Area and the Rest of Japan* (in Japanese). Kansai Economic Federation, Osaka, 1958.
77. JURÉEN, L. Long-term trends in food consumption: a multi-country study. *Econometrica*, vol. 24, no. 1, 1956, pp. 1–21.
78. KAHN, R. F. The relation of home investment to unemployment. *The Economic Journal*, vol. XLI, no. 162, 1931, pp. 173–98.
79. KALECKI, M. A macrodynamic theory of business cycles. *Econometrica*, vol. III, 1935, pp. 327–44.
80. KALECKI, M. *Studies in Economic Dynamics*. Allen and Unwin, London, 1943.
81. KALECKI, M. *Theory of Economic Dynamics*. Allen and Unwin, London, 1954.
82. KEYNES, John Maynard. *The General Theory of Employment, Interest and Money*. Macmillan, London, 1936.
83. KIRCHMAYER, Leon K. *Economic Control of Interconnected Systems*. John Wiley, New York; Chapman and Hall, London; 1959.
84. KIRSCHEN, E. S., and associates. *The Structure of European Economy in 1953*. O.E.E.C., Paris, 1959.
85. KLEIN, L. R., and H. RUBIN. A constant-utility index of the cost of living. *The Review of Economic Studies*, vol. XV (2), no. 38, 1947–48, pp. 84–7.
86. KOOPMANS, Tjalling C. Statistical estimation of simultaneous economic relations. *Journal of the American Statistical Association*, vol. XL, no. 232, pt. 1, 1945, pp. 448–66.
87. KOOPMANS, Tjalling C. Measurement without theory. *The Review of Economic Statistics*, vol. XXIX, no. 3, 1947, pp. 161–72.

88. KOOPMANS, Tjalling C. (editor). *Activity Analysis of Production and Allocation.* John Wiley, New York; Chapman and Hall, London; 1951.

89. KOOPMANS, Tjalling C., H. RUBIN and R. B. LEIPNIK. Measuring equation systems of dynamic economics. In *Statistical Inference in Dynamic Economic Models.* John Wiley, New York; Chapman and Hall, London; 1950.

90. LANCASTER, John Littlepage. *County Income Estimates for Seven South-Eastern States.* Bureau of Population and Economic Research, University of Virginia, Charlottesville, 1952.

91. LANGE, O., and F. M. TAYLOR. *On the Economic Theory of Socialism.* University of Minnesota Press, Minneapolis, 1938.

92. LEFEBER, Louis. *Allocation in Space.* Contributions to Economic Analysis, XIV. North-Holland Publishing Company, Amsterdam, 1958.

93. LEONTIEF, Wassily W. Quantitative input and output relations in the economic system of the United States. *The Review of Economic Statistics,* vol. XVIII, no. 3, 1936, pp. 105–25.

94. LEONTIEF, Wassily W. *The Structure of American Economy.* 1st edition *(1919–1929),* Harvard University Press, Cambridge, Mass., 1941; 2nd edition *(1919–1939),* Oxford University Press, New York, 1951.

95. LEONTIEF, Wassily. Inter-regional theory. In *Studies in the Structure of the American Economy.* Oxford University Press, New York, 1953.

96. LERNER, Abba P. *The Economics of Control.* Macmillan, New York, 1944.

97. LESLIE, P. H. On the use of matrices in certain population mathematics. *Biometrika,* vol. XXXIII, pt. III, 1945, pp. 183–212.

98. LESLIE, P. H. Some further notes on the use of matrices in population mathematics. *Biometrika,* vol. XXXV, pts. III and IV, 1948, pp. 213–45.

99. LOTKA, Alfred J. A contribution to the theory of self-renewing aggregates, with special reference to industrial replacement. *The Annals of Mathematical Statistics,* vol. X, 1939, pp. 1–25.

100. LUCE, R. Duncan, and Howard RAIFFA. *Games and Decisions.* John Wiley, New York, 1957.

101. MAHALANOBIS, P. C. Analysis of race mixture in Bengal. *Journal and Proceedings of the Asiatic Society of Bengal,* vol. XXIII (new series), no. 3, 1927, pp. 301–33.

102. MAHALANOBIS, P. C. On tests and measures of group divergence. *Journal of the Asiatic Society of Bengal,* vol. XXVI (new series), no. 4, 1930, pp. 541–88.

103. MAHALANOBIS, P. C. Anthropological observations on the Anglo-Indians of Calcutta. Statistical analysis of measurements of seven characters. *Records of the Indian Museum,* vol. XXIII, pt. III, 1940, pp. 151–87.

104. MAHALANOBIS, P. C. Discussion of the application of statistical methods in anthropometry. *Sankhyā,* vol. 4, pt. 4, 1940, pp. 594–8.

105. MAHALANOBIS, P. C. Some observations on the process of growth of national income. *Sankhyā,* vol. 12, pt. 4, 1953, pp. 307–12.

106. MEADE, J. E. *Consumers' Credits and Unemployment.* Oxford University Press, London, 1938.

107. MODIGLIANI, Franco, and Richard BRUMBERG. *Utility Analysis and Aggregate Consumption Functions: an Attempt at Integration.* Mimeographed.

108. MOORE, Frederick T., and James W. PETERSEN. Regional analysis: an inter-industry model of Utah. *The Review of Economics and Statistics,* vol. XXXVII, no. 4, 1955, pp. 368–83.

109. MOORE, Henry Ludwell. *Economic Cycles: their Law and Cause.* Macmillan, New York, 1914.

110. MOSER, C. A., and P. R. G. LAYARD. Planning the scale of higher education in Britain: some statistical problems. *Journal of the Royal Statistical Society, Series A*, vol. 127, pt. 4, 1964, pp. 473–526.

111. MOSER, C. A., and P. REDFERN. Education and manpower: some current research. In *Models for Decision*. The English Universities Press, London, 1965.

112. MOSES, Leon N. Inter-regional analysis. In *Harvard Economic Research Project: Report on Research for 1954* (mimeographed), pp. 163–226.

113. MOSES, Leon N. The stability of inter-regional trading patterns and input–output analysis. *The American Economic Review*, vol. XLV, no. 5, 1955, pp. 803–32.

114. MOSES, Leon N. An input–output, linear programming approach to inter-regional analysis. In *Harvard Economic Research Project: Report on Research for 1956–57* (mimeographed), pp. 22–50.

115. NERLOVE, Marc. The market demand for durable goods: a comment. *Econometrica*, vol. 28, no. 1, 1960, pp. 132–42.

116. NETHERLANDS, AMSTERDAM MUNICIPAL BUREAU OF STATISTICS. Stedelijke jaarekeningen van Amsterdam (Regional accounts for Amsterdam). *Kwartaalbericht van het Bureau van Statistick der Gemeente Amsterdam*, supplement, no. 4, 1953–4.

117. NEUMANN, John von, and Oskar MORGENSTERN. *Theory of Games and Economic Behavior*. Princeton University Press, second edition, 1947.

118. ORGANISATION FOR EUROPEAN ECONOMIC CO-OPERATION. *A Standardised System of National Accounts*. O.E.E.C., Paris, 1952; *1958 edition*, 1959.

119. PAIGE, Deborah, and Gottfried BOMBACH. *A Comparison of National Output and Productivity of the United Kingdom and the United States*. Joint study by O.E.E.C. and D.A.E., Cambridge. O.E.E.C., Paris, 1959.

120. PHILLIPS, A. W. Stabilisation policy in a closed economy. *The Economic Journal*, vol. LXIV, no. 254, 1954, pp. 290–323.

121. PHILLIPS, A. W. Stabilisation policy and time-forms of lagged responses. *The Economic Journal*, vol. LXVII, no. 266, 1957, pp. 265–77.

122. PHILLIPS, A. W. La cybernétique et le contrôle des systèmes économiques. *Cahiers de l'Institut de Science Economique Appliquée*, no. 72, 1958, pp. 41–8.

123. PITCHFORD, J. D. Growth and the elasticity of factor substitution. *The Economic Record*, vol. XXXVI, no. 76, 1960, pp. 491–504.

124. PONTRYAGIN, L. S., and others. *The Mathematical Theory of Optimal Processes*. Wiley (Interscience Publishers), New York, 1962.

125. PRAIS, S. J. Measuring social mobility. *Journal of the Royal Statistical Society, Series A*, vol. 118, pt. 1, 1955, pp. 56–66.

126. PRAIS, S. J. The formal theory of social mobility. *Population Studies*, vol. IX, no. 1, 1955, pp. 72–81.

127. PRAIS, S. J., and H. S. HOUTHAKKER. *The Analysis of Family Budgets*. Cambridge University Press, 1955.

128. PYATT, Graham. A measure of capital. *The Review of Economic Studies*, vol. XXX, no. 84, 1963, pp. 195–202.

129. PYATT, G. A production functional model. In *Econometric Analysis for National Economic Planning*. Butterworths, London, 1964.

130. RAIFFA, Howard, and Robert SCHLAIFER. *Applied Statistical Decision Theory*. Harvard Business School, Boston, 1961.

131. RAMSEY, F. P. A mathematical theory of saving. *The Economic Journal*, vol. XXXVIII, no. 152, 1928, pp. 543–59.

132. RAO, C. Radhakrishna. *Advanced Statistical Methods in Biometric Research*. John Wiley, New York; Chapman and Hall, London; 1952.

133. REDDAWAY, W. B. (editor). London and Cambridge Economic Bulletin. *The Times Review of Industry*, The Times Publishing Co., London, monthly. The Bulletin appears quarterly.

134. RILEY, Vera, and Saul I. GASS. *Linear Programming and Associated Techniques*. Johns Hopkins Press, Baltimore, 1958 (for the J.H.U. Operations Research Office).

135. ROOS, C. F., and V. von SZELISKI. Factors governing changes in domestic automobile demand. In *The Dynamics of Automobile Demand*. General Motors Corporation, New York, 1939.

136. ROY, S. N. A note on the distribution of the Studentised D^2-statistic. *Sankhyā*, vol. 4, pt. 3, 1939, pp. 373–80.

137. ROY, S. N. p-Statistics or some generalisations in analysis of variance appropriate to multivariate problems. *Sankhyā*, vol. 4, pt. 3, 1939, pp. 381–96.

138. SAMUELSON, Paul A. Some implications of 'linearity'. *The Review of Economic Studies*, vol. XV (2), no. 38, 1947–8, pp. 88–90.

139. SCHLAIFER, Robert. *Probability and Statistics for Business Decisions*. McGraw-Hill, New York, 1959.

140. SHUBIK, Martin. *Strategy and Market Structure*. John Wiley, New York, 1959.

141. SHUBIK, Martin, (editor). *Game Theory and Related Approaches to Social Behavior*. John Wiley, New York, 1964.

142. SIMON, Herbert A. A formal theory of interaction in social groups. In *Models of Man*. John Wiley, New York; Chapman and Hall, London; 1957.

143. SLATER, Lucy Joan. A dynamic programming process. *The Computer Journal*, vol. 7, no. 1, 1964, pp. 36–9.

144. SMITH, Walter L. Renewal theory and its ramifications. *Journal of the Royal Statistical Society*, Series B, vol. XX, no. 2, 1958, pp. 243–302.

145. SOMMERVILLE, D. M. Y. *An Introduction to the Geometry of N Dimensions*. Republished by Dover Publications, New York, 1958.

146. STONE, Richard. On the interdependence of blocks of transactions. *Supplement to the Journal of the Royal Statistical Society*, vol. IX, nos. 1–2, 1947, pp. 1–45.

147. STONE, Richard. The theory of games. *The Economic Journal*, vol. LVIII, no. 230, 1948, pp. 185–201.

148. STONE, Richard. Linear expenditure systems and demand analysis: an application to the pattern of British demand. *The Economic Journal*, vol. LXIV, no. 255, 1954, pp. 511–27.

149. STONE, Richard. Transaction models with an example based on the British national accounts. *Boletin del Banco Central de Venezuela*, April 1955 (in Spanish). *Accounting Research*, vol. VI, no. 3, 1955, pp. 1–24.

150. STONE, Richard. *Quantity and Price Indexes in National Accounts*. O.E.E.C., Paris, 1956.

151. STONE, Richard. *Input–Output and National Accounts*. O.E.E.C., Paris, 1961.

152. STONE, Richard. An econometric model of growth: the British economy in ten years time. *Discovery*, vol. XXII, no. 5, 1961, pp. 216–9.

153. STONE, Richard. Models for demand projections. In *Essays on Econometrics and Planning*. Pergamon Press; Statistical Publishing Society, Calcutta; 1965.

154. STONE, Richard, and Giovanna. *National Income and Expenditure*. Seventh edition. Bowes and Bowes, London, 1964.

155. STONE, Richard, J. AITCHISON and J. A. C. BROWN. Some estimation problems in demand analysis. *The Incorporated Statistician*, vol. 5 no. 4, 1955, pp. 165–77.

156. STONE, Richard, and J. A. C. BROWN. A long-term growth model for the British economy. In *Europe's Future in Figures*. North-Holland Publishing Co., Amsterdam, 1962.

157. STONE, Richard, and J. A. C. BROWN. Output and investment for exponential growth in consumption. *The Review of Economic Studies*, vol. XXIX, no. 80, 1962, pp. 241–5.

158. STONE, Richard, and Alan BROWN. Behavioural and technical change in economic models. In *Problems in Economic Development*. Macmillan, London, 1965.

159. STONE, Richard, Alan BROWN, Graham PYATT and Colin LEICESTER. *Economic Growth and Manpower*. British Association for Commercial and Industrial Education, London, 1963.

160. STONE, Richard, Alan BROWN and D. A. ROWE. Demand analysis and projections for Britain: 1900–1970. A study in method. In *Europe's Future Consumption*. North-Holland Publishing Co., Amsterdam, 1964.

161. STONE, Richard, D. G. CHAMPERNOWNE and J. E. MEADE. The precision of national income estimates. *The Review of Economic Studies*, vol. IX, no. 2, 1942, pp. 111–25.

162. STONE, Richard, and Giovanna CROFT-MURRAY. *Social Accounting and Economic Models*. Bowes and Bowes, London, 1959.

163. STONE, Richard, and D. A. ROWE. Aggregate consumption and investment functions for the household sector considered in the light of British experience. *Nationaløkonomisk Tidsskrift*, vol. 94, pts. 1 and 2, 1956, pp. 1–32.

164. STONE, Richard, and D. A. ROWE. The market demand for durable goods. *Econometrica*, vol. 25, no. 3, 1957, pp. 423–43.

165. STONE, Richard, and D. A. ROWE. Dynamic demand functions: some econometric results. *The Economic Journal*, vol. LXVIII, no. 270, 1958, pp. 256–70.

166. STONE, Richard, and D. A. ROWE. The durability of consumers' durable goods. *Econometrica*, vol. 28, no. 2, 1960, pp. 407–16.

167. STONE, Richard, and D. A. ROWE. A post-war expenditure function. *The Manchester School of Economic and Social Studies*, vol. XXX, no. 2, 1962, pp. 187–201.

168. STONE, Richard, and others. *The Measurement of Consumers' Expenditure and Behaviour in the United Kingdom, 1920–1938*, vol. I. Cambridge University Press, 1954.

169. THEIL, Henri. *Optimal Decision Rules for Government and Industry*. North-Holland Publishing Co., Amsterdam, 1964.

170. THURSTONE, L. L. *Multiple-Factor Analysis*. The University of Chicago Press, 1947.

171. TODHUNTER, I. *A History of the Mathematical Theory of Probability*. Macmillan, Cambridge and London, 1865.

172. TUSTIN, Arnold. *The Mechanism of Economic Systems*. Heinemann, London, 1953.

173. TUSTIN, Arnold. Economic regulation through control-system engineering. *Impact of Science on Society*, vol. IV, no. 2, 1953, pp. 83–110.

174. U.K., BOARD OF TRADE and CENTRAL STATISTICAL OFFICE. *Input–Output Tables for the United Kingdom, 1954*. Studies in Official Statistics, No. 8. H.M.S.O., London, 1961.

175. U.K., CENTRAL STATISTICAL OFFICE. *Economic Trends*. H.M.S.O., London, monthly.

176. U.K., CENTRAL STATISTICAL OFFICE. *National Income and Expenditure*. H.M.S.O., London, annually.

177. U.K., COMMITTEE ON HIGHER EDUCATION (ROBBINS COM-MITTEE). *Higher Education*. Report, Cmnd. 2154. H.M.S.O., London, 1963.
178. U.K., COMMITTEE ON HIGHER EDUCATION (ROBBINS COM-MITTEE). *Higher Education*. Appendix I, Cmnd. 2154-I. H.M.S.O., London, 1963.
179. U.K. GOVERNMENT. *Employment Policy*. Cmd. 6527. H.M.S.O., London, 1944.
180. U.K. MINISTRY OF AGRICULTURE, FISHERIES AND FOOD. *Domestic Food Consumption and Expenditure*. Annual Reports of the National Food Survey Committee. H.M.S.O., London, annually since 1952.
181. UNITED NATIONS, STATISTICAL OFFICE. *A System of National Accounts and Supporting Tables*. U.N., New York, 1953; *Revision 1*, 1960.
182. U.S. CONFERENCE ON RESEARCH IN INCOME AND WEALTH. *Regional Income*. Studies in Income and Wealth, vol. 21, Princeton University Press, 1957.
183. U.S., DEPARTMENT OF COMMERCE. *Personal Income by States Since 1929*. U.S. Government Printing Office, Washington, 1956.
184. VAJDA, S. *Readings in Linear Programming*. Pitman, London, 1958.
185. VINING, Rutledge. Methodological issues in quantitative economics. *The Review of Economics and Statistics*, vol. XXXI, no. 2, 1949, pp. 77–94.
186. WALRAS, Léon. *Eléments d'Economie Politique Pure*. Definitive edition, R. Pichon et R. Durand-Auzias, Paris and F. Rouge, Lausanne, 1926 and later reprints. English translation by William Jaffé; George Allen and Unwin, London, 1954.
187. WARNTZ, William. *Toward a Geography of Price: a Study in Geo-econo-metrics*. University of Pennsylvania Press, Philadelphia, 1959.
188. WHITE, Harrison C. *An Anatomy of Kinship*. Prentice-Hall, New Jersey, 1963.
189. WOLD, Herman. A synthesis of pure demand analysis. *Skandinavisk Aktuarie-tidskrift*, vol. XXVI, 1943, pp. 85–118 and pp. 222–63, and vol. XXVII, 1944, pp. 69–120.
190. WOLD, Herman, and Lars JURÉEN. *Demand Analysis*. John Wiley, New York; Almqvist and Wiksell, Stockholm; 1953.
191. WORKING, E. J. What do statistical 'demand curves' show? *The Quarterly Journal of Economics*, vol. XLI, 1927, pp. 212–35.
192. YZEREN, J. van. Three methods of comparing the purchasing power of currencies. *Statistical Studies*, no. 7, Netherlands Central Bureau of Statistics, 1956.